Berliner ökophysiologische

und phytomedizinische Schriften

Hrsg. von Christian Ulrichs und Carmen Büttner

Lebenswissenschaftliche Fakultät

Humboldt-Universität zu Berlin

Band 53

Hrsg. von

Dr. Maria Landgraf

Humboldt-Universität zu Berlin

Detection of viruses of the virome affecting birches (*Betula* sp.) - ''Case studies Europe-wide''

D I S S E R T A T I O N

zur Erlangung des akademischen Grades

Doctor rerum agriculturarum

(Dr. rer. agr.)

eingereicht an der
Lebenswissenschaftlichen Fakultät
der Humboldt-Universität zu Berlin

von

Elisha Bright Opoku

geboren am 22.06.1988, in Accra-Ghana

Präsidentin
der Humboldt-Universität zu Berlin
Prof. Dr. Julia von Blumenthal

Dekan der Lebenswissenschaftlichen Fakultät
der Humboldt-Universität zu Berlin
Prof. Dr. Dr. Christian Ulrichs

Gutachter/innen
1. Prof. Dr. Carmen Büttner
2. Prof. Dr. Susanne Jochner-Oette

Tag der mündlichen Prüfung: 24.01.2023

Bibliografische Information der Deutschen Nationalbibliothek

Die Deutsche Nationalbibliothek verzeichnet diese Publikation in der Deutschen Nationalbibliografie; detaillierte bibliographische Daten sind im Internet über http://dnb.d-nb.de abrufbar.

1. Aufl. - Göttingen: Cuvillier, 2023

Zugl.: Berlin, Univ., Diss., 2023

ISBN 978-3-7369-7753-2

eISBN 978-3-7369-6753-3

Table of content

1 Introduction

Viruses are small obligate intracellular parasites consisting of either DNA or RNA genome surrounded by coat of proteins known as capsid (Gelderblom, 1996). The DNA or RNA genome may consist of single stranded (ss) or double stranded (ds) with complex shape mostly spherical or tubular (Louten, 2016). Different replication strategies are required for the various types of genome. For replication of RNA viruses, at least three types of RNA must be synthesized i.e. the genome, a copy of genome and mRNAs using virally encoded RNA-dependent-RNA-polymerase (RdRp) (Payne, 2017). Viruses need living host cells for replication because they have no ribosomal machinery hence the need to rely on protein synthesis present in their host cells (Walsh *et al.*, 2011).

Viruses are known to have emerged from evolution and ecological processes due to interaction with their host over some period of time (Elena *et al.*, 2014). For plant virus emergence, they are mostly mediated by changes in agricultural practices and long-distance transportation of plant materials (Rojas and Gilbertson, 2008). They can have either positive or negative impacts on their hosts in several ways. Roossinck, in 2015 reported that, virus may be either beneficial or harmful leading to the death of plants. This means, there can be a mutualistic relationship between virus and host depending on the environments and vice versa. Viruses interact with their host plant and the environment thereby exchanging genetic information between hosts (McLeish *et al.*, 2019).

Although viruses are capable of virtually infecting different species of cultivated and wild plants, their host range varies from narrow to wide depending on individual type of viruses (Gergerich *et al.*, 2006). The type and severity of the host reactions to virus infections are variable and depends on the virus strains, sources of infection, crop genotypes as well as environmental conditions (Hull, 2014). For example, the *Cucumber mosaic virus* is known to have the widest host range worldwide among plant viruses while *Citrus tristeza* infects only a few species in the genus *Citrus* (Gergerich *et al.*, 2006).

Some plant viruses are vector transmissible e.g. fungi, nematodes, aphids and other insects whiles others are transmitted via pollen, seeds and seedlings, grafting, roots and mechanical transmission (Büttner *et al.*, 2022). For cultivated woody plants, they are transmitted via vegetative propagation depending on specific properties of the virus (Büttner *et al.*,2022). Depending on the mode of transmission, different strategies for preventing the spread of plant viruses need to be taken into consideration and has been described in chapter 6.

With an estimated annual economic impact of over $30 billion, about 50% of the known plant viruses have been reported to be a contributing factor in agricultural crop losses globally (Sastry and Zitter, 2014). Much research has been done on viruses infecting fruit trees such as grapevine, mango, pome fruit, banana and citrus (Umer *et al.*, 2019). Nevertheless, there is little or no knowledge on viruses infecting forest and urban trees (Büttner *et al.*, 2013). Hence there is a need to focus on urban green space since they have immense impact on the health of our populations (Konijnendijk *et al.*, 2013). Although viral diseases in trees are widespread, there is a lack of extensive data on viruses infecting forest and urban trees (Büttner *et al.*, 2013).

Forest pathology previously focussed traditionally on insect damage and fungal diseases (Linnakoski and Forbes, 2019) without considering viruses affecting forest and urban trees. Trees for instance, in urban green areas are subject to great stress and often more short-lived than their conspecifics in a natural environment (Roloff, 2013). Not only abiotic stress factors such as nutrient deficiency, pollutants in both air and water, contaminants in soil, fine dust, road salt and ozone influence the health of tree (Czaja *et al.*, 2020), but also many biotic factors (e.g. soil microorganisms, insects, bacteria, nematodes, viruses etc.). Viral diseases, which are able to increase the predisposition to other pathogenic factors, have been found on degenerating deciduous trees in cities for over a decade now (Büttner *et al.*, 2013). For instance, Cooper and Massalski (1984) observed that, complex symptoms of changes in colour and shape of leaves from some deciduous trees such as *Betula* sp., the loss of branches and crowns are associated with virus infections.

The birch (*Betula* sp.) which is deciduous pioneer tree was selected as a model tree due to the fact that suspected viral symptoms have been observed on them for some years now in urban green (Division of phytomedicine -HUB). *Betula* belonging to the family *Betulaceae* is one of the most relevant broad-leaved trees in Northern and Eastern Europe which contributes to the biodiversity of coniferous forest (Hynynen, *et al.*, 2010). The common species in Europe are the *Betula pendula* Roth and *pubescens* Ehrh. (Ashburner and McAllister, 2013). Even though both tree species (*B. pendula* Roth. and *B. pubescens* Ehrh.) are naturally distributed across Europe to central Siberia, *B. pubescens* is widely distributed in the north and eastern regions and grows further north than other deciduous trees whereas the *B. pendula* can reach southern regions such as Iberian Peninsula, South Italy and Greece (Beck *et al.*, 2016). The birch is an important tree species in the cities, because they provide habitats for numerous native animal species.

Birch is of considerable value to insectivores, cavity nesters, seed-eaters and field-layer species and has moderately diverse bird community compared to that of mixed deciduous woods (Patterson, 1993). According to (Heydemann; *Der Forst- und Holzwirt, p. 536*, Hanstein 1984), there are a total of 164 types of insect in birch trees, including 11 weevils, 27 longhorn beetles, 10 bark beetles, 9 owl butterflies. *B. pendula* do survive under dry weather conditions with decreasing growth performance whiles the *B. pubescens* are normally not suitable for locations with high temperatures (Asche *et al.*, 2007). Both tree species requires a minimum length of growing season of 110 days with relatively significant characteristics shown in Tab.1.

Tab. 1 Birch wood cultivation with different climatic growth conditions (Asche, *et al.*, 2007)

	Minimum length of growing season in days	Trophy level	Very dry	Dry	Moderately dry	Moderately fresh	Fresh	Very fresh	Ground fresh	Basic damp	Damp	Wet	Water logged	Alternative damp	moderate alternating damp	Alternative dry
Betula pubescens	110	3-5														
Betula pendula	110	1-5														

Tree species suitable for the location with good growth performance

Tree species appropriate to the location with decreasing growth performance and/or increasing risk

Tree species not suitable for the site

With a stock volume ranging from 0.5% to 15% of all hardwood standing, depending on the region, *Betula* species play a key role in the forests of Western Europe (Dubois *et al.*, 2020). In Germany, they contribute about 5% stock volume of hardwood (Hynynen *et al.*, 2010). The *Betula* sp. which has largely been neglected appears to be alternative species for reforestation because they are more adaptable to future climate change and can generate valuable products for future market (Hemery *et al.*, 2010; Dubois *et al.*, 2020).

Betula sp. can withstand climate changes and tolerate climate fluctuations well if the site requirements of the respective birch species are taken into consideration. Planting birch trees as a pioneer tree also makes economic sense, especially in urban re-cultivation areas. The demands of *Betula* species, which are particularly advantageous in extreme locations such as in the city, are also derived from the distribution area. Wrong site location or selection sites are a major detrimental factor in keeping trees in the city for a long term. Since the *Betula* sp. are sensitive to summer heat (Teskey *et al.*, 2015), the location for planting in the city must be carefully considered. This is because both species as well as their hybrids differ, their site location and the choice of species in relation to the birch species play an essential role for growth after transplanting shock and thus for the long-term stability of birch trees in cities.

The general question arises whether *Betula* sp. suffer from viruses in cities more than country side due to city climate or whether trees cultivated for growth in the cities suffer due to poor management practices in the nurseries. Even though *Betula* sp. can adapt to extreme weather conditions, biotic factors such as viruses may influence the vulnerability of trees after being predisposed by abiotic factors. According to reports, agricultural systems, such as monocrops with low genetic diversity and high plant density, which are more susceptible to pathogens; global trade in plant materials; effects of climate change on hosts and vectors; and capacity for rapid evolution and adaptation are factors contributing to the emergence of viruses (Anderson *et al.*, 2004; Jones, 2009; Elena *et al.*, 2014).

The aim of this thesis is to collect data and generate more knowledge on the frequency and distribution of the viral infections in *Betula* sp. In the case of urban trees there is no information on how long birch population fight this infection already. The question arises whether it's an old or a new problem since they are now being detected in some parts of Europe. The *Betula* sp. population seems to be resisting those viruses already for long time and can combat these viruses under optimal conditions. From my personal observation, natural grown sites and forests had less symptoms compared to *Betula* trees in urban green space. This is probably because in urban areas, there are no optimal conditions and is engulfed with more viruses from all over the world due to artificial ecosystems with man-made diversity.

Again, other plant viruses which are not yet detected in *Betula* can trigger infection. In addition to fungal diseases, viruses are of particular importance, which can manifest themselves predominantly on trees in unsuitable locations under the increasingly extreme climatic conditions in urban green space. Externally, these viral diseases are primarily characterized by the loss of branches, causing possible death of trees which is common in fruit trees but has not yet been shown in birches. The loss of branches influences stability and road safety due to dead branches which causes an increase in management and maintenance cost (Brookes, 2007). The distributions of viruses on road trees are being investigated in this study. Although green urban space is considered as a major positive and cost-effective means of improving the quality of the environment (Navarrete-Hernandez and Laffan, 2019), the complex ecology of urban areas is constantly being disrupted locally by the emergence of new type of species and concentration of chemicals into specific habitats due to human activity (Douglas, 2008).

All these factors mixed with wrong site location, seedlings contaminated with viruses, inadequate attention as well as plant shock affects the resistance of trees to pathogens for years and can often lead to the early death of the trees. In order to fully understand the ecological effects, more research is required. For instance, studies are missing, if virus infected birch trees are more attractive for insects like the birch bark beetle (*Scolytus ratzeburgii*) leading to a fast decline and death of the trees (Tidow, 2020). Plant virus induced attraction by changes in leaf appearance and coloration is known from field crops since a long time (Jones and Barbetti, 2012), hence the need to also focus on urban trees. This effect might be even worse if abiotic stresses are intervening. Strict measures for city planning including protection and examining young tree seedlings diseases such as viruses will be essential in managing these and other difficulties as the world's urban areas increases.

In our basic study the focus was on birch plus trees in seed plantation, as well as younger and old trees in urban green space. In a follow up study, seeds and seedlings from the tree nurseries are to be examined in order to narrow down the ways in which seedlings can be contaminated with viruses. As mentioned before, naturally grown locations of the birch compared with the locations of street trees have noticeable less virus suspected symptoms. This is an indication that the seed-borne viral diseases get into our street trees with the tree nursery material (Cooper, 1979). On the other hand, the means of transmission for most viruses is not yet understood and seems to differ in forest and urban green due to the differences in the ecosystem and the human impact. One of our hypotheses is that, birch populations grown under natural conditions can cope better with viral diseases because only the fittest and /or, virus-free seedlings survives in a given environment. The situation is different with the cultivation of seedlings in the nursery sector, where all seedlings (including infected viral seedlings which would not survive in nature) are processed and cultivated under optimal conditions as much as possible. The transmission routes of pathogens, especially viruses, must be recognized and interrupted, especially if they are spread by human activities during cultivation of trees.

As viruses cannot be treated with pesticides or medication, the major focus in a management concept can only be on early diagnostic of viruses and removal of virus infected trees to interrupt transmission. In this way we can show where the viral diseases come from and open up the possibility of finding and eliminating the sources of infection. Paths for sustainable planting and maintenance of trees free from pathogenic viruses- or infected trees are to be developed. Research has to answer the question if the particular viruses of host plant in a certain virome have a clear pathogenicity and how they contribute to the decline.

It is certainly very difficult to prevent viral infection from trees in case of vector transmitted viruses or if trees are connected to each other by root contacts and thus mixing their microbiome. To protect young trees as long as possible from infection by pathogenic viruses, prophylactic measures should be considered since that might give them a better chance to overcome plant shock and other abiotic stresses. A very hygienic environment for the production of the virus free-trees is by selecting virus-free source material for budding and grafting. The status of viruses in forest trees and the recommendations to combat them can therefore only be determined after basic research in order to get information on mode of transmission of viruses, host range, geographical distribution and other factors.

A concept for dealing with virus-infected trees is also preceded by the identification and detection of the causative pathogens and knowledge of their epidemiology and interactions with the host. On the other hand, trees cannot be ignored as sources of viruses in our ecosystem (Harris and Hill, 2021), as viruses often have a broad range of host plants and can be transmitted to cultivated plants. Based on knowledge on viruses (e.g. *Cacao swollen shoot virus, Apple mosaic virus, Cherry leaf roll virus, Grapevine leafroll-associated virus* etc.) infecting fruit growing trees, it is obvious that viruses can also be found in deciduous trees. These can make the trees sick and must be counted among the pathogens, for example if they reduce the plant quality, the fruit quality, seed germination, the biomass and the longevity of trees. Ultimately, they lead to the degeneration of the entire tree or tree population and thus to long-term damage to the ecosystem.

A severe outbreak of *Citrus tristeza virus* in citrus growing regions during last century is reported to have destroyed almost 100 million trees (Moreno *et al.*, 2008). There are also reports on loss of yield efficiency of peach infected with mild isolate of *Plum pox virus* even though tree produced slightly more fruits of smaller size than non-infected trees. A mixed infection of *Plum pox virus* (PPV), *Prune dwarf virus* (PDV), and *Prunus necrotic ringspot virus* (PNRSV) has been reported to reduce growth by exhibiting bark canker, trunk malformation, and tree mortality in some peach cultivars and in reduction of growth rate of seedlings by 2.9 to 69.1% (Nemeth, 1992). Trees on the other hand seem to combat viral infections by developing a kind of equilibrium with its virome. My personal observation in some older individual birch trees (about 80 years) from urban green space from Berlin were found to be infected with several viruses including symptoms but still survive because of growing in a more natural environment.

Symptoms can be named in two ways, whether they are symptoms of virus, diseases, of other infectious diseases or mere physiological or genetic disorders (Bos, L. 1977). Virus induced symptoms are confused with nutrient deficiency, drought or ozone damage or even fungal infection. Often one finds chlorosis (yellowing), subsequent necrosis (drying out, apoptosis), premature leaf-shedding and weak branches as clear symptoms of a viral disease in *Betula* sp. (Büttner *et al.*, 2022). The expression and pattern in the leaves depend on the respective viral disease or viral combination. The hypothesis has been generated that the diversity in symptom pattern (i.e. line pattern, chlorotic line pattern, leaf mosaic, small leaves, intercostal chlorosis, leaf necrosis) as shown in (Fig. 1) reflects the diversity of the viruses present in the *Betula* trees. As the correlation of symptoms and viral infection is not examined yet for the mixed infections, it is unknown if the complexity of the virome is the cause of the variability of symptoms.

Fig. 1 Diversity of viral *Betula* leaf symptoms; Line pattern (1), chlorotic line pattern (2), leaf mosaic (3), small leaf (4), intercostal chlorosis (5), leaf variegation (6), line pattern (7), leaf necrosis (8). (Division of phytomedicine - Humboldt University of Berlin (HUB).

Many of the viruses detected in *Betula* (Tab.2) have so far been completely unexplored and their impact on the domestic ecosystem is difficult to assess. It is therefore particularly important to identify these, obtain new knowledge about the pathogenicity of these viruses and to be able to diagnose the pathogens among them in a targeted manner. Obvious virus-suspect symptoms on newly planted trees only show up after a few years or under harmful environmental conditions. For tree nurseries or seed plantations, it is very difficult and expensive to find out about a tree infection after a couple of years cultivating diseased trees.

Especially if those viruses belong to pathogens affecting fruit quality or tree health. The trees must be removed from field and cannot be sold without the consequence of losing the quality standards of the nursery and resulting in immense financial losses. Therefore, the diagnostic of pathogenic viruses as early as possible in the production process is of utmost importance. However, the longevity of trees is the basis for the long-term establishment of many organisms such as pathogens in the tree ecosystem (Lindenmayer *et al.*, 2017). If the longevity is reduced from the outset during planting by pathogenic viruses contained in the planting material (young trees or seedlings), a long-term existence of trees in the ecosystem cannot be established. For this reason, targeted virus diagnostics in the plant material is particularly important, which has not yet been carried out due to a lack of knowledge and research into pathogenic viruses of deciduous trees. This is a different situation in production of fruit trees such as grapes or berries, where the production is already based on an efficient virus management.

Tab. 2 Viruses detected in host plant (*Betula* sp.)

Host plant	Detected virus	Reference
Betula sp.	*Apple mosaic virus*	Bandte *et al.*, 2009; Cooper and Massalski, 1984, Gotlieb and Berbee 1973.
Betula sp.	*Arabis mosaic virus*	Bandte *et al.*, 2009; Polák and Zieglerová, 1997; Polák and Procházková, 1996; Hardcastle and Gotlieb, 1980; Gotlieb and Berbee, 1973.
Betula sp.	*Birch leaf roll virus*	Rumbou *et al.*, 2018.
Betula sp.	*Cherry leaf roll virus*	Büttner *et al.*, 2011; von Bargen *et al.*, 2009; Jalkanen *et al.*, 2007; Jones *et al.*, 1990; Nienhaus and Castello, 1989; Cooper and Atkinson, 1975.
Betula sp.	*Tobacco necrosis virus*	Cooper and Massalski, 1984.
Betula sp.	*Tomato ringspot virus*	Cooper and Massalski, 1984.

In field crop production, the loss of plants and the economic impact due to viruses is about 50% before virus management (Jones *et al.*, 2019). Unfortunately, it is very difficult to show an economic loss due to viruses in the production of deciduous trees. Mainly this has something to do with the long-life span of trees compared to field crops, which require long term experiments and research project financing. Financing of this need deep interest of the society in the topic.

As long as tree diseases are hardly recognized, the research on tree viruses is focussing on obvious symptoms as represented by the Birch leaf roll disease (BLRD). The improvement of the sustainability of tree planting, virus diagnostics, control management of viral diseases and prophylactic hygiene measures in maintenance are to be mentioned as necessary steps towards an effective virus management. Basic knowledge on the biology of the viral pathogen is the prerequisite to gain information on how to interrupt life cycle of viruses or to block viral transmission in a given population. Epidemiology and pathogenicity of the newly discovered viruses in birch as well as in other hosts, host specificity, life cycle, mode of transmission, host plant range and phylogeny, are totally unknown and have to be investigated especially for so far unknown viruses.

Knowledge of the viral status of gene banks, nurseries, forest stands, and trees in urban green space, such as roadside trees, can add supporting information on the biology of the viruses. This knowledge can be used for developing control plans and diagnostic tools for tree management. Many so far unknown viruses such as *Birch leaf roll associated virus* (BLRaV) have been identified by application of High-throughput sequencing (HTS) in deciduous trees (Rumbou *et al.*, 2018). BLRaV is associated with severe symptoms in birch leaves (i.e. chlorosis, leaf rolling, vein banding and necrosis) and shall be further investigated in this study to underline its contribution to the Birch leaf roll disease (BLRD). Therefore, one chapter is particularly reviewing the knowledge on BLRaV and other badnaviruses.

The hypothesis rises that; viral diseases in woody trees begins from "mother" trees and tree nursery stock that are grown with inadequate selection and under poor hygienic condition. Since the tree seedlings are not certified as virus-free unlike the fruit growing trees, there is a possibility of increased incidence of viral diseases, which are then transferred to the urban green areas and contributes to the decline of city climate stressed trees. One important way viruses are transmitted is from seeds and seedlings onto urban green which weakens trees in the long term. Another way of increasing viral infection in urban green is linked to the observation that, birch populations that have developed under natural conditions can cope better with viral diseases. This is of great importance to clarify whether the investigated viral diseases get into our street trees by inadequate seedling selection. One can assume other ways of transmission depending on the individual properties of the viral species. For instance, a transmission by pollen or decaying leaves or vectors has to be considered as well. Furthermore, there are indications that plant viruses could be a contributing factor to the decline of *Betula* sp. depending on the choice of species, health status of planting material as well as location.

The distribution of viruses affecting *Betula* trees in seed plantation might give information if viral infections starts from seed plantation. Leaf samples showing virus suspected symptoms from streets of bigger cities (Berlin) can give information on the dimension of viral diseases already distributed in the urban green space. In order to fill a huge gap of knowledge about viral diseases, such studies can monitor how widespread are viral symptoms in *Betula* trees in the city.

As the correlation of symptoms and viral infection is not examined yet, it is unknown if the complexity of the virome is the cause of the variability of symptoms. Based on the investigations, better decisions can be made in order to plant and maintain healthy *Betula* trees in the urban green space in the future. To this end, diagnostic procedures must be developed and refined. Only then can the importance of the different viruses currently being discovered in the *Betula* trees and their damage to the ecosystem be recognized. In our ecosystem's, organisms are clustered into parasites and symbionts (Rynkiewicz *et al.*, 2015). This separation applies for viruses as well, but beneficial viruses are even more difficult to determine because an infection and replication in a host cell by a virus implies per se a negative impact on the tree even if the beneficial impact is of a secondary mode for instance by protecting the host from a more aggressive viral strain (Roossinck, 2015). Those beneficial viruses might exist in a complex mixture of viruses as we find it in *Betula* and they might drive the tree towards equilibrium with its microbiome. This is supported by the infection by viruses in older trees, which find ways to combat it.

The situation in mixed viral infected trees is highly complex but reflexes the true situation in nature, where we find lots of viruses associated with natural populations of plants (Harris and Hill, 2021). With less information known on plant viruses affecting deciduous woody plants, the HTS has become a reliable tool for further diagnostic and discovery of viral agents in woody hosts which is more reliable (Massart *et al.*, 2017; Roossinck *et al.*, 2015) like particle isolation, enrichment and bioassay followed by molecular characterization. The use of HTS hasproven as good as or better than bioassays in detecting viruses and viroids in woody plants (Rott *et al.*, 2017; Al Rwahnih *et al.*, 2015). For example, about 21 new viruses infecting forest trees have been discovered within the last five years (Rumbou *et al.*, 2021) via HTS including the genus badnavirus. Several steps such as; nucleic acid extraction and virus sequences enrichment, library preparation, automated sequencing, data analysis are needed to be taken into consideration when applying HTS technologies.

Building on these successful stories, the application of HTS techniques for routine virus detection has gained momentum. One of the viruses detected is the BLRaV and there are hints on the potential presence of more badnavirus species in the datasets (Rumbou *et al.*, 2018; Köpke, 2019). BLRaV affecting *Betula* sp. has been successfully detected on the RNA level by RT-PCR; hence consequently there is a need to obtain data from the DNA level. The amplification of DNA is an important step which involves digestion of RNA. By database analysis using bioinformatics, the so far known sequence information on the DNA level needs to be evaluated and potential virus information needs to be extracted for further investigation. This promises to improve the knowledge on the existence of BLRaV or relatives within the plant birch genome for later characterization of the virus and its pathogenicity.

The National centre for biotechnology information (NCBI) provides information on the genome of *Betula* sp. which is used as reference sequence to find more badnavirus related sequence information in the genome of birch applying the following tools. A search program provided by the NCBI known as the Basic local alignment search tool (BLAST) is used for comparing primary biological sequence information such as nucleotide sequences of DNA or RNA as well as amino-acid sequences of protein (NCBI, 2021). Although virus discovery using HTS is a powerful tool, there is need for verification using alternative methods. Current bioinformatics pipelines can discover a virus-like sequence but cannot distinguish the two forms; not allowing for a clear view of the significance findings and therefore additional steps are needed to identify the virus form (Shahid *et al.* 2021).

For example, some dsDNA (badnaviruses) viruses can be found in two forms: the first one is when the virus integrates into the plant genome and does not necessarily cause a disease; and the other, episomal, when the virus actively replicates in the cell and may cause disease (Harper *et al.*, 2002). Likewise, while not all viruses are known to integrate into the plant genome, evidence of the presence of a circular genome via rolling circle amplification is also needed for circular DNA viruses. A database search can be conducted to answer whether there is more undetermined sequence information on Badnaviruses existing in published sequence data from birch genomes following the hypothesis that BLRaV as a badnavirus/pararetrovirus capable of integration into the host genome.

The general objective of this study was to identify, characterize to a certain degree, and determine the distribution of potential pathogenic viruses affecting birches with a special focus on BLRaV.

The specific objectives of this project concern:

1. The mode of transmission of viruses relevant in urban tree management and to show that BLRaV forms are involved in transmissible infection.
2. The seed production trees are already affected by birch viruses.
3. A review on the viral complex in birch investigated so far.
4. The correlation of symptoms towards viral infection and more unknown viruses affecting birches in potential mixed infection.
5. Viral signatures and viruses with similarity to BLRaV birch databases.

2 Materials and methods

Information on chemical and protocols are listed in Appendix 11.2 and Appendix 11.3 repectively.

2.1 Survey of selected *Betula* stands -Berlin- Germany

Experimental design on a systematic survey of virus infected *Betula* sp. was once a year conducted between May and June (2015, 2016 and 2017), to determine the appearance and distribution of viruses infecting *Betula* sp. from eight districts in Berlin. Monitoring of selected *Betula* stands in Berlin was done based on cadastre registration provided by the "Grünflächenamt". Declined birches from different stands were selected based on the data from the "Grünflächenamt". Leaf samples were collected in the following areas; Wilhelm Gericke Straße (Reinickendorf), Mauerpark Birkenwäldchen (Pankow), Tiergarten (Global stone project)(Mitte Tiergarten), Park am Gleisdreieck (Friedrichshain-Kreuzberg), Treptower Park(Treptow-Köpenick), Tempelhofer Feld (Tempelhof Schöneberg), Unter den Birken (Treptow-Köpenick), verlängerte Rathenaustraße (Treptow- Köpenick), Schlangenbader Straße (Charlottenburg-Wilmersdorf), Lentzeallee (Steglitz-Zehlendorf), Rabindranath-Tagore-Straße (Treptow- Köpenick) and Waltersdorfer Chaussee (Neukölln) (Appendix 11.3; Table A20 for specific locations). A lift truck was provided by cooperation partners from Berlin parks department (Bezirksamt, Grünflächenamt) for easy access to crown of trees. Based on suggestions from Grünflächenamt, declining birches (*Betula pendula* and *Betula pubescens*) showing obvious symptoms like weak appearance, dead branches were selected (Fig. 2 and Fig. 3).

Fig. 2 Examples of *Betula* sp. leaf symptoms observed in streets of Berlin (June, 2017). Vein banding and ringspot, Blankenstein Park (1=E54074), leaf spot in Blankeinstein Park (2=E54074), necrosis and mottling, Park am Gleisdreieck (3 =E54068), vein banding with necrosis, Blankenstein Park (4 =E54076), vein banding and necrosis, Schlangenbader Straße (5= E54030), loss of crown and decline of branches, Lentzeallee (Steglitz-Zehlendorf) (6).

Trees were identified by visual rating and with the aid of Berlin tree cadastre. Birch trees were scored for the presence or absence of virus suspected symptoms and distinguished from insect bite, fungi, parasites by identification of mosaic-like leaf patterns, chlorotic ring spots, leaf mottling, degeneration or loss of crown as well as leaf deformation (Landgraf *et al.*, 2016). By photographic documentation of different stages of leaf symptoms were determined and categorized into different classes of symptoms (Appendix 11.4; Figure A1). Symptomatic leaves were collected from 72 birch trees (*Betula* sp.). Each leaf sample was numbered, and stored at -80°C for further molecular analysis.

Fig. 3 Examples of *Betula* sp. leaf symptoms observed in different streets of Berlin (June, 2017). Ringspot, Blankenstein Park (7=E54074), leaf spot and vein banding, Blankeinstein Park (8=E54075), necrosis and mottling, Blankeinstein Park (9=E54075), Chloritc spot, Blankenstein Park (10=E54075), chlorotic spot and necrosis in Blankeinstein Park (11= E54076), vein banding, Schlangenbader Straße (12 =E54031), necrosis, Schlangenbader Straße (13 =E54031), ringspots Schlangenbader Straße (14 =E54031), Loss of tree crown Schlangenbader Straße (15 =E54031), Weak branches Schlangenbader Straße (16 =E54031).

2.2 *Betula* sp. from Wildberg/Nagold seed plantation-Germany

A survey in a selected seed plantation with *Betula* sp. was carried out together with the DFG-project partner Prof. Dr. Susanne Jochner-Oette from Katholischen Universität Eichstätt-Ingolstadt and Mr. Thomas Ebinger from Staatsklenge Nagold. Leaf samples from 71 *Betula* sp. trees were collected in June, 2019 by the team of Phytomedicine division -HUB. The *Betula* sp. in Wildberg/Nagold seed plantation were mostly propagated and cultivated by cloning (Appendix 11.3; Table A22). Specific characteristics typical for so called 'plus- trees' were used for seed production and tree breeding according to "Forstvermehrungsgut-Zulassungsverordnung" (FoVZV, 2002). In all, 24 different birch clones were investigated withleaf sample assigned with E-number. Some of the viral disease symptoms are shown in (Fig. 10).

2.3 Selected *Betula* stands -Rovaniemi-Finland

The birch leaf symptoms observed in some areas in Finland are more intense and easier to identify (oral communication by von Bargen). To gain more information on the viral distribution in affected birches, 26 *Betula* trees with viral suspected leaf symptoms from different areas in Rovaniemi (Finland) were kindly provided by Division of Phytomedicine-HUB in June, 2017 for further investigations. Samples were taken from upper parts of birch trees with kind support from Dr. Risto Jalkanen (METLA) in some areas in Finland (Appendix 11.3; Table A21). Leaf symptoms were categorized into groups of leaf roll, intercostal chlorosis, ringspots and necrosis as described by Landgraf *et al.*, 2016; and listed in Appendix 11.4; Figure A1.

2.4 RNA isolation from plant materials (Boom *et al.*, 1990)

Total RNA was extracted from all symptomatic and non-symptomatic *Betula* leaf samples collected from Berlin (Germany-72 *Betula sp.*), Wildberg/Nagold (Germany-71 *Betula sp.*), and Rovaniemi (Finland-26 *Betula sp.*) according to Boom *et al.*, 1990. Fresh *Betula* leaves (0.3g) were ground homogenously in universal bags (BIOREBA) filled with 3000 µl grinding buffer solution. 500 µl of this homogenous mixture were then transferred to two separate 2ml Eppendorf tubes containing 100 µl of 10% SDS solution, vortex using (Janke and Kunkel IKA VF2) and incubated in a Thermomixer (MB-102 Mixing Block, BIOER) at 70°C for 10 minutes at a speed of 800 rotations per minute (rpm).

Each homogenous mixture in the 2ml Eppendorf tube was left on ice for 5 minutes and then centrifuged using (Eppendorf 5424R) at 15,500 x g at 4°C for 10 minutes. 300 µl supernatant from each tube was collected and added to two separate 1.5ml Eppendorf tube with each tube containing a mixture of (300 µl sodium iodide solution, 150 µl of 96% ethanol (non-denatured) and 25 µl of silica solution. After careful mixing (vortex for 10 seconds), the reaction tubes were incubated at room temperature for 10 minutes. Within these 10 minutes, each reaction tube was inverted twice for optimal binding of nucleic acids to the silica particles with remaining proteins denatured by SDS solution. The silica solution was centrifuged at a temperature of 4°C, 6000 x g (Eppendorf 5810R) for 1 minute and the supernatant discarded. The silica particles containing the viral RNA were repeatedly washed and centrifuged twice using 500 µl washing buffer at a temperature of 4°C, 6000 x g for 1 minute using the centrifuge device (Eppendorf 5810R) and all supernatant were discarded. With a little pellet of silica at the bottom of the 1.5ml, each tube was inverted and incubated at room temperature for 10 minutes until all washing buffer were removed and dried. 100 µl of DEPC H_2O was added to silica pellet and then incubated at 70°C with a Thermomixer (MB-102 Mixing Block, BIOER) for 4mins, at 800rpm.

The silica pellet with DEPC H_2O was centrifuged for 5 minutes, 15.500 x g at 4°C (Eppendorf 5810R). After centrifugation, 100 µl supernatant containing the RNA were pipetted into a 1.5ml Eppendorf tube containing a mixture of 500 µl of 96% ice-cold ethanol (-20°C) and 8 µl LiCl (8M). This was stored overnight at -20°C. After precipitation, the mixture was centrifuged for 20 minutes at 18.000 x g at 4°C using centrifuge (Eppendorf 5810R) and the supernatant was carefully discarded. 500 µl of washing buffer was then added and centrifuged 18.000 x g at 4°C (Eppendorf 5810R) for 5minutes. The supernatant was carefully discarded again. Using a speed vacuum (SC110, Savant), the total RNA samples in 1.5ml Eppendorf tubes were dried for 2 minutes. 30 µl DEPC H_2O was then transferred into each dried 1.5ml Eppendorf tubes containing RNA samples and allowed to cool on ice for 15-30 minutes. This was followed by measurement of the quality and quantity of RNA concentrations from leaf samples.

Grinding Buffer	6 M Guanidine hydrochloride
	0.2 M Na-Acetate pH 5.2
	25mM EDTA
	1 M Potassium acetate,
	2.5 % (w/v) Polyvinylpyrrolidone (PVP-40)

Sodiumdodecylsulfate (SDS) solution	10 % (w/v) SDS

Sodium iodide solution	6 M Sodium iodide(NaI)
	0.15 M Sodium sulphite (Na_2SO_3)

Silica suspension	50% (w/v) SiO_2 with DEPC H_2O (pH 2.0)

Washing buffer	10 mM Tris-Cl pH 7.5
	0.5 mM EDTA
	50 mM Sodium chloride (NaCl)
	50 % (v/v) 96 % Ethanol (non- denatured)

2.4.1 Quantitative measurement of RNA Concentration

The concentration of total RNA was assessed by measuring the absorption at 260 nm and 280 nm using a NanoDrop One/One[C] micro volume UV/VIS Spectrophotometer (Thermo Fisher Scientific Company). 2µl of each total RNA extract were pipetted onto the surface of the spectrophotometer's light path, observed on a screen and recorded at the absorption wavelength 260/280nm and 260/230 nm respectively. For the ratio of 260/280nm, a value of 1.85 to 1.88 is assumed to be pure for DNA and a value of 2.1 is assumed to be pure for RNA. The 260/230 ratio is used to indicate the presence of unwanted organic compounds such as phenol, guanidine hydrochloride and guanidine thiocyanate. For the ratio of 260/230, a value of 1.8 to 2.2 is optimal for RNA.

A lower ratio of RNA sample is an indication of the presence of contaminants that absorb strongly at or near 280 nm. Generally acceptable 260/230 ratios are in the range of 2.0 to 2.2. Values higher than this may indicate contamination with the compounds which has already been mentioned. Concentration of each RNA sample were calculated according to the law of Lambert and Beer whose algorithm was included in the software package of the NanoDrop One/One[C] micro volume UV/VIS Spectrophotometer and was output in Nano gram per microliter (ng per µl).

For each RNA, the total concentration is shown in (Appendix 11.3; Table A20, Table A21 and Table A22). The Nano-Drop read provides an indication of the presence of RNA but the integrity of the RNA has to be confirmed in further experiments.

2.4.2 Qualitative evaluation of RNA

The first step for evaluation of quality of RNA was to prepare a gel. The gel chamber plates, gel comb and other gel materials was first denatured by using (10 % (w/v) SDS solution). Further steps has been described in chapter 2.8. The RNA concentration was measured qualitatively by running RNA sample on a denaturing 1% gel agarose and staining with SERVA DNA stain. The ribosomal RNA which consists of (28S rRNA and 18S rRNA, 16S rRNA and 23S rRNA) was displayed on a screen after a successful running of gel electrophoresis. The quality of the RNA results depends on the ratio of 28S rRNA and 18S rRNA. Two bands without smear corresponding to 28S rRNA and 18S RNA represents a good quality of RNA. A ribosomal ratio significantly lower than 2:1 is a sign of RNA degradation. The probability that RNases act as a mediator between 28S rRNA and 18S rRNA band is higher due to the length. Hence 28S rRNA band with optimal RNA extraction would more clearly visible than 18S rRNA band.

2.5 Reverse transcription (cDNA synthesis)

Total RNA extracted from all *Betula* leaves were synthesized to complementary DNA (cDNA). Two different approaches were used in the transcription of the RNA to cDNA. The first approach was the use of a reverse transcriptase enzyme pMMLV-RTase (Division of Phytomedicine-HUB) originally encoded by Moloney Murine Leukemia Virus. Based on the concentration of the RNA (1200-1500ng), about 2.0 µl to 5.0 µl were pipetted and transferred into 0.5ml Eppendorf tube containing 2.0 µl random hexamer (50µM). DEPC H$_2$O was added to the tube containing RNA extracts and random hexamer primer to a volume of 10 µl, mixed by vortex (Janke and Kunkel IKA VF2). This was incubated at 70°C (BIOMETRA-T personal) for 5 minutes and immediately cooled on ice for 5 minutes. A master mix of 10 µl was prepared (Tab. 3) and transferred into the cooled RT component (RNA samples, random hexamer, pMMLV-RTase and DEPC H$_2$O).The complete reverse transcription component was mixed by vortex for 30 seconds and incubated for 120 minutes at 37°C. After 120 minutes of incubation, the reaction components were immediately incubated 5 minutes at 85°C (BIOMETRA-T personal) to deactivate the enzymes completely. The protocol for the first method (Tab. 3 and Tab. 4) was used in the investigation of *Betula sp.* from Berlin (Germany) and Rovaniemi (Finland).

Tab. 3 Volume and reverse transcription reaction components.

Volume (1X reaction)	RT Component
2-5 µg	RNA
2 µl	Primer (50 µM) Random Hexamer
Up to 10 µl	H_2O DEPC

Tab. 4 Volume and reverse transcription reaction using pMMLV-RTase.

Volume (1X reaction)	RT Component
2 µl	dNTPs (10mM each)
4 µl	5 x reaction buffer MMLV-Rtase (Phytomedizin)
0.5 µl	RNAse Inhibitor Ribolock (40 u/µl)
2 µl	pMMLV-RTase (Phytomedizin)
1.5 µl	DEPC H_2O
10 µl	Total volume

10 µl of reaction mix were added to 10 µl of RT component (RNA, random hexamer and DEPC H_2O), incubated at 37°C for 120 minutes and deactivation of the enzyme at 85°C for 5 minutes.

For the second method, total RNA was transcribed to cDNA using the reverse transcription components; total RNA of 1200-1500ng depending on the concentration of total RNA, random hexamer (2.0 µl),1 µl dNTP and Milli Q H_2O adding up to a volume of 14.5 µl (Tab. 5). The reverse transcription components were incubated at 65°C (BIOMETRA-T personal) for 5 minutes and cooled on ice for 5 minutes (Tab. 5).

Tab. 5 Reverse transcription reaction.

Volume (1 reaction)	RT Component
1200-1500ng	Total RNA
2 µl	Random hexamer (50 pmol/µl)
1 µl	dNTP mix (Each 10mM)
x µl	Milli Q H_2O
14.5 µl	Total volume

Reverse transcription components were incubated 65°C for 5 minutes.

Tab. 6 Reverse transcription reaction using Maxima H Minus reverse transcriptase.

Volume (1 reaction)	RT Component
4 µl	5 x RT-Buffer (Thermo Scientific)
0.5 µl	RNAse Inhibitor Ribolock (40u/µl)(Thermo Scientific EO 0382)
0.5 µl	Maxima H Minus Reverse Transcriptase (Thermo Scientific)
0.5 µl	Milli Q H_2O
5.5 µl	Total volume

Tab. 7 Reverse transcription PCR program.

Temperature	Time
50°C	30 min
85°C	5 min
10°C	∞

The enzyme reverse transcriptase Maxima H Minus (Thermo Fisher Scientific) (0.5 µl), 5 X RT buffer (0.5 µl) and RiboLock RNase Inhibitor (0.5 µl) was pipetted into 2ml Eppendorf tube and vortexed for 10 seconds. 5.5 µl was transferred into Eppendorf tube containing the total RNA together with RT components (Tab. 5). The samples were then incubated using the (BIOMETRA-T personal) at different temperatures of 50° C (30 minutes) and 85°C (5 minutes) (Tab. 7). For *Betula* samples from Wildberg/Nagold, Maxima H Minus reverse transcriptase was used for cDNA production.

2.5.1 Quality control of cDNA

To analyse the applicability of cDNA in a downstream PCR, the cDNA was tested in a model PCR targeting a universal plant mitochondrial nad5 gene transcript. Kato *et al.,* 1995 characterised a part of the mitochondrial DNA from apple and described the existence of the nad5 gene (NADH dehydrogenase subunit 5) containing the exons a and b, separated by an 848 base pairs intron (Menzel *et al.*, 2002). A master mix of 2.5µl 10 X PCR buffer without $MgCl_2$, 0.5 µl of dNTP, 0.25 µl of both forward and reverse primer nad5 (Menzel *et al.*,2002), 2µl $MgCl_2$ (25mM),1 µl of Taq polymerase at a ratio of 1:20 was prepared and transferred into 0.5 µl PCR tubes filled with 1 µl of cDNA (Tab. 8). Using a PCR program (Tab.9) with denaturation of 94°C, annealing temperature of 62°C and elongation temperature of 72°C at a cycle of 35 X within an estimated time, all cDNA samples were tested with the protocol as shown in (Tab. 8 and Tab. 9).

Tab. 8 Volume and Reverse transcription components used for nad5 gene as a control of the integrity of RNA.

Volume (µl)	RT Component
1	cDNA
2.5	10 X PCR buffer without $MgCl_2$ (Phytomedizin)
0.5	dNTPs (10mM each)
0.25	nad5 Foward (50 µM)
0.25	nad5 Reverse (50 µM)
1	Taq Polymerase (5u/µl Phytomedizin) (1:20)
2	$MgCl_2$ (25 mM)
17.5	MilliQ H_2O
25	**Total volume**

Tab. 9 Nad5 PCR program.

	Temperature	Time	Number of cycles
Primer denaturation	95°C	2 min	1 x
Denaturation	94°C	30 s	
Primer annealing	**62°C**	30 s	35 x
Elongation	72°C	60 s	
Final elongation	72°C	5 min	1 x
Hold	10°C	∞	

The PCR products were separated in a 1.5% agarose gel at 80V for about 30 minutes and stained by using fluorescence dyes (SERVA DNA Stain G) under UV light.

2.6 Oligonucleotides (Primers)

All oligonucleotides used for RT-PCR detection of plant viruses were ordered from Biolegio Company (Netherlands). Each lyophilized primer was diluted up to a stock concentration of 500 µM in DEPC distilled H_2O. All primers were further diluted with DEPC distilled H_2O to a final concentration of 50 µM needed for laboratory work.

Tab. 10 The primers resulted in the production of specific amplification size by RT-PCR.

Virus-organism	Primer name	Primer Sequence 5'-3'	Target	Product size	Reference
BLRaV	BadnaSG Forward	ACGAAGAAGGATACCTCACGG	Coat protein	242bp	Pack, 2016 (BSc.)
	BadnaSG Reverse	CCAGTGTCTAGTACCGCCC			
	Badna K Forward	GCAGGCACAGAACAATGG		300bp	Pack, 2016 (BSc.)
	Badna K Reverse	GGCATTACTAGCCATTCGTA			
BiCV	Carla F	GGCACTTACGTACAGGAGT	Coat protein	197bp	Pack, 2016 (BSc.)
	Carla R	GCGGAAAAGGGGCTTAGATA			
CLRV	CLRV-CP350F	CATRGAGAAGCTAAATTTCTCTCT	RNA2-Coat Protein	627bp	Langer et al., 2016 Demiral,2014 (MSc.)
	CLRV-CP977R	ACTCMACCCTATCAAARTATAYCA			
	CLRV-CP188 F	TGTCTGTRAATATTATGGCTGGTAC	RNA2 coat protein coding region	170bp	von Bargen et al., 2009
	CLRV-CP350 R	AGAGAGAAATTTAGCTTCTCYATG			
ApMV	ApMV -CP1739R	GGTGGTAACTCACTCGTTAT	Coat protein	204bp	Langer,2016 (Unpublished)
	ApMV -CP1535F	GTAATCCGAAAGGTCCGAAT			
nad 5(Plant Mitochondria genome)	nad 5 Forward	GATGCTTCTTGGGGCTTCTTGTT	plant NADH subunit 5 gene	181bp	Menzel et al., 2002
	nad 5 Reverse	CTCCAGTCACCAACATTGGCATAA			

2.6.1 Primer design for new viral signatures

Primers were derived from the HTS contigs which showed a clear similarity to a reference virus by so far unknown genera in birch using Primer-Blast (Rozen and Skaletsky, 2000) and primer Express (Applied Biosystems Primer Express™ Software v3.0.1 License). The Primer-BLAST program combines information on hybridization efficiently with stereo chemical information on the selected oligonucleotide. Contigs which were larger and showed a high coverage, E-value close to zero and high similarly (>80%) to a reference sequence were used to design new primers. The potential function of the area chosen was derived from the similarity to reference viruses after BLASTx.

Using the BioEdit program (version 7.1.3.0) (Hall, 1999) and ClustalW algorithm (Thompson et al., 1994), the virus-suspected contigs were aligned with the known nucleotide sequences of reference strains. In primer design, areas conserved in the unknown sequence and reference strain are preferred with the aim of detecting one genome region from as many strains as possible or primers on the genus level. Only specific oligonucleotides with both biophysical and stereo chemical optimal conditions were selected for use by means of RT-PCR (3' self-complementary in primer BLAST close to zero).

The online program tool OligoCalc (Kibbe, 2007): Oligonucleotide Properties Calculator and Primer- Express together with the Primer- BLAST were used to evaluate the stereospecifity of the primers. Hairpin formation and self-complementary structures were avoided by checking the stereochemistry in OligoCalc. The melting temperature was always calculated by the neighbour-joining method (Ye *et al.*, 2012). Due to the limited information on sequence diversity and sometimes-small similarity to reference strains, most of the new designed primers are probably strain specific. The primers resulted in the production of different amplification sizeof PCR products (Tab. 11).

Tab. 11 The primers listed resulted in the production of different amplification size of PCR products.

Virus-organism	Primer name	Primer Sequence 5'-3'	Target	Product size	Reference
Potential Benyvirus	Beny1856 Forward Beny1856 Reverse	AGGCTGCTAAGGTTTTCCG AGACAACTCACGGGACGAAC	replication-associated protein RNA1	254bp	This study
Potential Capillovirus	Capillo Forward Capillo Reverse	CCATCTTCTCTTGCGCGT GGGGGCAAAGAAGTGACGA	coat protein	469bp	This study
Potential Caulimovirus	Caulimo410A Forward Caulimo410A Reverse	GCTTTGAGAATTGCACTCCAG TAAGATTCTCCGGCACTGGG	polyprotein OFR5 Polymerase	162bp	This study
Potential Deltapartitivirus	Deltapartite324 Forward Deltapartite324 Reverse	TGGTGAAGTACCAGAACGCA AGTGGCGCTGCTGAAAG	RNA1	579bp	This study

2.7 Polymerase chain reaction (PCR)

The polymerase chain reaction (PCR) is a molecular diagnostic tool used in specific amplification and detection of targeted DNA sequences from a complex of nucleic acid. A combination of specific primers, deoxyribose nucleotide triphosphate (dNTP) and thermostable Taq DNA polymerase leads to the formation of a new DNA molecule starting from 3' end of the primer. PCR components (listed in Appendix 11.3: Table A2) were used in the amplification of target sequence through repeated cycles of denaturation, annealing and elongation at specific temperatures. The temperatures were determined by the sequence of the primers and the optimal working temperature of the enzyme leading to an exponential increase in DNA molecules.

For each reaction, a master mix containing 2.5 µl of 10 X PCR buffer solution with MgCl$_2$ 0.5 µl dNTPs, 0.5 µl from forward and reverse primers (50µM), 2.0 µl MgCl$_2$ (25mM), 16µl Milliq H$_2$0 and 1 µl of Taq polymerase (Phytomedizin 5µl 1:20) was pipetted into 2ml Eppendorf tube. 23 µl of master mix solution was transferred into a tube containing 2 µl of prepared cDNA, vortexed for a short time and transferred into a Thermal cycler (BIOMETRA-T personal). Due to pipetting errors, 1µl H$_2$O and a reserve was always added to the master mix. Based on different primer designs, *Birch leaf roll associated virus* (BLRaV), *Birch carla virus* (BiCV), *Cherry leaf roll virus* (CLRV), *Apple mosaic virus* (ApMV) with specific targets of amplification as well as the unknown viral signatures was investigated.

2.7.1 RT-PCR method for detection of CLRV

For CLRV detection by RT-PCR, different primers (Appendix 11.3; Table A1) were designed based on available sequence information at NCBI. The coat protein area of binding for CLRV with primers (CLRV CP350F/977R with accession numbers (S63537.1 and MK402282.1) are known to be specific. The other primer combination of CLRV-CP 188F/CP350R) with the accession numbers (MN399681.1 and MK402282.1) are known to be specific. Both primers (CLRV CP350F/997R and CLRV-CP 188F/CP350R) were used in amplification of all cDNA samples at an amplicon size of 627bp and 170bp respectively. With a well-designed PCR program, each cDNA samples were amplified with an annealing temperature of **53°C**. The chemical components reaction and PCR program is listed in (Appendix 11.3; Table A3 and Table A4).

2.7.2 Detection of BLRaV by RT-PCR

Using the molecular diagnostic tool RT-PCR, the primer pair BadnaSG (SG corresponds to Schwarzer Grund) listed in (Appendix 11.3; Table A1) proposed by Pack, 2016 were used in amplification of cDNA from *Betula* leaf samples at an amplicon size of 242bp. Another primer pair published in the same work BadnaK (where K corresponds to Korsika) was used to investigate each of the cDNA samples resulting in an amplification product of 300bp. The PCR components and PCR program are listed in (Appendix 11.3; Table A5, Table A6 and Table A7).

2.7.3 RT- PCR method for BiCV detection

A full genome of BiCV with the accession number (MH536506.1) affecting birches has been published (Rumbou *et al.*, 2020) and primers used in this work were derived from its sequence as described before. The primer set combination of Carla K (K corresponds to *Korsica*) derived from strains from Corsica (France) (Pack, 2016), were used in the investigation of all samples from Berlin, Wildberg/Nagold and Rovaniemi. For RT-PCR aiming to confirm the presence of BiCV, the primers (Appendix 11.3; Table A1) 5'-GGCACTTACGTACAGGAGT-3' (F) and 5'-GCGGAAAAGGGGCTTAGATA-3' (R) were applied targeting the conserved coat protein region at an amplicon size of 197bp. Chemical components and PCR program are listed in (Appendix 11.3; Table A8 and Table A9).

2.7.4 Detection of ApMV by RT-PCR

A primer set 5'-GGTGGTAACTCACTCGTTAT-3' (ApMV-CP1535 Forward) and 5'-GTAATCCGAAAGGTCCGAAT-3' (ApMV -CP1739 Reverse) (Langer *et al.*, 2016 unpublished) were used for PCR detection of ApMV using reaction components (see Appendix 11.3; Table A10) and PCR program (see Appendix 11.3; Table A11) at specific amplicon size of 204bp.

2.7.5 Detection of partial Benyvirus signature by RT-PCR

Using the molecular diagnostic tool RT-PCR, the primer pairs named Beny 1856 Forward and Beny 1856Reverse designed from this study were validated in this work as shown in (Appendix 11.3; Table A12 and Table A13) at the amplicon size of 254bp by gel electrophoresis.

2.7.6 Birch Capillovirus-like sequence

A first PCR primer combination targeting coat protein (5'-CCATCTTCTCTTGCGCGT-3') (F) and (5'- GGGGGCAAAGAAGTGACGA-3') (R) was designed (see Tab. 11) for further validation (Appendix 11.3; Table A16 and Table A17). It amplified a specific fragment of the capillovirus-like sequence at product size of 469bp and was confirmed by Sanger sequencing.

2.7.7 Caulimoviridae-like sequences in birch different from BLRaV

Based on HTS data from 2018, RT-PCR primers were designed as described in chapter 2.6.1. and provided by Division of phytomedicine-HUB for further validation. Existence of the corresponding RNAs in the original leaves was confirmed by Sanger sequencing of RT-PCR products for contig410 and contig0108. For the contig410 forward primer Caulimo410AFW 5'-GCTTTGAGAATTGCACTCCAG-3' and reverse primer Caulimo410ARW 5'-TAAGATTCTCCGGCACTGGG-3' derived from the polyprotein OFR5, Polymerase were provided and resulted in 149bp product at 53°C in standard RT-PCR. For contig0108 the primer combination (Caulimo108FW 5'-TGCATCGTCATCAGCAACTC-3' and Caulimo108RW 5'-CGTTGTCAGGGCTTGTGTTG-3') targeting the reverse transcriptase with a final RT-PCR product of 186 bp was designed. Only primers targeting contig 410 were chosen for the following screening analysis due to their high specificity (Appendix 11.3; Table A14 and Table A15).

2.7.8 Signatures of Deltapartitivirus-like sequence in birch

The primers Deltapartiti324F 5'-TGGTGAAGTACCAGAACGCA-3' and Deltapartiti324R 5'-AGTGGCGCTGCTGAAAG-3' were designed based on HTS data from 2014 and provided by Division of phytomedicine-HUB. Samples from Berlin and Wildberg/Nagold were tested for Deltatpartitivirus using the chemical components and PCR-program provided in (Appendix 11.3; Table A18 and Table A19).

2.8 Agarose gel electrophoresis

The process of gel electrophoresis is applying an electrical current to nucleic acids in gel matrix. Negatively charged DNA or RNA moves from the cathode to the anode inducing size separation of the molecules depending on the pore size of the gel matrix. The velocity of nucleic acid molecules in agarose gels depends on two factors: the size of the molecule and concentration of the agarose defining the pore size of the gel matrix (Lee *et al.*, 2012). DNA is negatively charged at neutral pH and thus migrates to the anode. A gel was prepared by weighing 1.0g -1.5 g of agarose (Biozym) corresponding to the concentration mentioned in PCR protocol (Appendix 11.3; Table A20), diluted in 100ml of (1X TBE buffer) and warmed using a microwave for 2 to 3 minutes until all solute of agarose were dissolved. 6.0μl -10.0 μl of DNA stain Clear G (1:10) from SERVA (Serva Company) which is a fluorescent dye in a ratio of 1: 100,000 were then added to agarose solution after cooling down for 5 minutes. The nucleic acids are stained with an intercalating fluorescent dye and therefore visible under UV light.

The agarose solution was poured into a gel tray attached with a comb and allowed to cool down for 30 minutes. After gel polymerization, the gel tray containing polymerized gel was then transferred into a gel electrophoresis chamber which contained 1 X TBE buffer. The comb was gently removed from the gel tray and levelled up with 1XTBE buffer to cover the gel. The gel electrophoresis was used to analyse the quality of RNA, as well as PCR products from all *Betula* leaves. For a PCR protocol, 6 μl of PCR product mixed with 2μl loading dye was transferred onto each gel well. 3.0 μl to 3.5μl standard molecular (1Kb or 50bp Gene Ruler DNA ladder) (Thermo Fisher Scientific Company) was pipetted on one of the precast gel well as a reference marker. A current between 70 to 80 volt (Ms major science, MP300V) was programmed to run through the nucleic acid fragment for 30 to 45 minutes depending on fragment size and the PCR protocol. The gel was removed after the required minutes was reached and transferred onto a gel documentation system (E-Box from Vilber smart Imaging Company). Different PCR product fragments were observed under ultraviolet light and and printed out. These fragment were then compared with the Gene Ruler marker which consist of 50bp as well as 1Kb depending on the size of the PCR product as weel as RNA fragments being analysed.

5x Loading buffer   **1x TBE running buffer, 1liter**

100 mM EDTA, pH 8.0

50% (v/v) Glycerol   90mM Tris-Borate
                                         2mM EDTA (pH 8.3)
0.025% (w/v) Bromophenol blue

0.025% (w/v) Xylene cyanol

2.9 Sanger Sequencing of PCR products

After successful amplification of the required PCR products, they were cleaned up using the MSB Spin PCRapace or Invisorb Fragment CleanUp-Kit (STRATEC Molecular) purified to keep it free from inhibitors such as dNTPs, dyes, or salts. The PCR product was mixed with 250µl binding buffer and pipetted onto a spin filter column and centrifuged for two minutes at 11,000 xg. In this step the DNA bind to the membrane and separate from the remaining components of the PCR mixture. The spin filter was transferred to a sterile 1.5ml reaction tube. 20µl elution buffer was transferred onto the spin filter and incubated at room temperature for one minute. It was then centrifuged further for one minute at 11,000 xg in order to remove the DNA. The spin filter was then discarded and the sample was sent out for external sequencing at the company Macrogen. Using the PCR product specific primers, with target of interests, sequencing was carried out.

2.10 Double Antibody Sandwich - Enzyme-linked immunosorbent assay (DAS-ELISA)

To confirm the existence of CLRV by a second independent method showing the level of viral proteins in the birch leaf samples, double antibody sandwich enzyme-linked immunosorbent assay (DAS-ELISA) was applied. The serological diagnostic tool DAS-ELISA was used for the detection of *Cherry leaf roll virus* (CLRV) in affected birches (Breuhahn *et al.*, 2013). CLRV isolates used as positive controls were derived from cherry and elderberry produced by BIOREBA AG Company (Switzerland). The two different isolate-specific polyclonal antibody sets against the cherry strain of CLRV (CLRV-ch) and elderberry (CLRV-e), respectively, were provided by BIOREBA Company. As the antibodies are polyclonal, they can generate non-specific results. As they are derived from specific isolates, they cannot recognise all isolates from CLRV so they complement each other. In other words, the antibody CLRV-ch can react with cherry isolates but not with isolates from elderberry whiles CLRV-e reacts with elderberry but not with cherry. Eleven (11) symptomatic leaves were taken from different parts of a specific tree (TT SEAL 0407510 with E-54094) for DAS-ELISA experiment.

In addition to that, Different leaves from 6 birch seedlings (407553BpenMO291B, 407554BpenMO291D, 407555BpenMO291F, 407556BpenMO291H, 407569BpenMO345B confirmed to be infected by CLRV (Cutler, 2016 personal communication) and *E-54095* by Reverse Transcriptase polymerase chain reaction (RT-PCR). All leaf materials were collected from Lentzeallee 55/57 (Berlin-Germany).

With three replications for each leaf sample as well as positive and negative controls, 6300 µl of coating buffer was transferred onto two different petri dishes. 6 µl of antibody coating IgG (IgG 1:1000, CLRV-e) was added to the first petri dish containing the coating buffer. 6 µl of antibody coating IgG (IgG 1:1000, CLRV-c) was transfered onto the second petri dish containing coating buffer. After careful mixing, 100 µl coating buffer from each petri dish was transferred into each well on two different ELISA plates (Nunc MaxiSorp F96 microtiter plates). Each ELISA plates were labeled ELISA plate -c and ELISA plate-e repectively. Each ELISA plate were then covered with Parafilm and incubated at 30°C for 4 hours. After coating, the antibodies to the surface of the ELISA plates, the plates were washed three times using 200 µl of washing buffer.

Washing buffer was allowed to stay in wells for 2 minutes before discarded by carefully inverting the ELISA plate. After last wash, each well was dried by emptying it on using paper towels. Using the extraction buffer each birch leaf sample was ground homogenously together with extraction buffer (400-800 µl), in (BIOREBA) bags. 100 µl of plant extracts were pipetted into each ELISA plate well.

Plates were covered with Parafilm and incubated at 4°C overnight. In this step, the viral coat proteins from plant extracts bind with antibodies. The plates were carefully washed three times with 200 µl washing buffer per well.

Conjugate buffer was prepared by pipetting 6 µl of conjugate IgG w/AP (alkaline phosphatase) for each (CLRV-e and CLRV-ch) respectively in a petri dish containing 6300 µl of conjugate buffer and carefully mixed.100 µl was added to each well, covered with Parafilm and incubated at 30°C for 5 hours. The secondary antibody connected to the alkaline phosphatase forms an antigen -antibody complex after incubating for 5 hours.

The plates were washed three times using washing buffer (200 µl for each well) and emptied on paper towels after incubation. 100 µl of substrate buffer (containing substrate of AP) was pipetted into each well and plates were incubated at a room temperature (20°C-25°C in the dark) for 30-120 minutes. The samples were measured by ELISA reader (MULTISKAN FC, Thermo Scientific), using a filter wavelength (405nm). The colour reaction towards yellow indicates positive results and the negative results without any colour.

This can be evaluated visually or photo-metrically. The thresh hold was calculated using the formula (mean value + 3 x standard deviation + 10%) (BIOREBA).

Extraction buffer	2% (w/v) PVP 40 20 mM Tris Base 137 mM NaCl 2.7 KCl 1 mM $MgCl_2$ x $6H_2O$ 0.05% (w/v) Tween 20 pH 7.3-7.5
Coating buffer	159 mg Na_2CO_3 293 mg $NaHCO_3$ 20 mg NaN_3 pH 9.5-9.7
Washing buffer	137 mM NaCl 1.5 mM KH_2PO_4 8.1 mM Na_2HPO_4 2.7 mM KCl 0.05% (w/v) Tween20 pH 7.2-7.6
Conjugate buffer	2% (w/v) PVP 40 20 mM Tris Base 137 mM NaCl 2.7 KCl 1 mM $MgCl_2$ x $6H_2O$ 0.05% (w/v) Tween 20 0.2% (w/v) pH 7.3-7.5
Substrate buffer	1 M Diethanolamin 1 mg/ml pNPP (para-nitrophenyl-phosphate) pH9.7-9.9
ELISA reagents	*Cherry leaf roll virus*-elderberry strain (CLRV-e) Coating IgG (Lot No. 021537) IgG conjugated w/AP (Lot No. 031537) Positive control (Lot No. 065985) *Cherry leaf roll virus*-cherry strain (CLRV-e) Coating IgG (Lot No. 011134) IgG conjugated w/AP (lot no. 021134) Positive control (Lot No.055511)

2.11 Bioassays

Symptomatic *Betula* leaves with confirmed BLRaV- badnavirus infection (RT-PCR) were used for bioassay experiment. Seeds from indicator plants (*Chenopodium quinoa, Nicotiana tabacum . var. Samsun, Nicotiana benthamiana and Chenopodium amaranticolor*) were sown four weeks before inoculation.

Fig. 4 *Betula* leaf samples with BLRaV infections used for bioassay experiments. E54094 (1) and E54095 (2), inoculated indicator plants (3).

0.3g of each *Betula* leaf sample (E54094 and E54095) with BLRaV infection, inoculation buffer (Na$_2$SO$_3$ at pH 7.0, 0.2% (w/v) and celite) were homogeneously grinded in (BIOREBA) extraction bags. A sterilized pin was used to induce a small hole in the primary leaf surface of all indicator plants and then inoculated to two plants of each indicator species. The inoculum was applied by gently rubbing onto the leaf surface of all indicator plants. After inoculation, the leaves of all indicator plants were washed with water to remove excess sap and abrasive.

Inoculated plants were daily observed for symptoms development for three weeks to four weeks. Bioassay plants were then harvested after three weeks. Parts of the harvested leaf samples were further inoculated onto new indicator plants (*C. amaranticolor and C. quinoa)* and the remaining leaves were stored in -20°C.

Leaf samples from all indicator plants were subjected to RNA extraction by Boom *et al.* 1990 as described in chapter 2.4. RNA was transcribed into cDNA using random hexamers and pMMLV reverse transcriptase (as described in chapter 2.5). This was followed by RT-PCR for nad5 transcript. Molecular biological techniques (RT-PCR) was used to test BLRaV with a primer combination of Badna SG (Forward and Reverse) expecting a PCR product size of 242base pairs (as described in chapter 2.7.2). The experiment was repeated in 2018 using the same protocol with a different enzyme Revert Aid (Thermofisher Scientific Company) for cDNA production and followed with RT-PCR experiment.

2.12 High-throughput sequencing (HTS) diagnostics

The high-throughput sequencincing diagnostics involves several steps which includes; sampling of leaf materials, purification, precipitation and concentration of RNA, removal of rRNA from totalRNA which is then followed by cDNA generation.

2.12.1 Sampling

The leaf material was collected from Wildberg/Nagold birch seed plantation in June 2019. The two trees (E58128 and E58145) showed significant viral symptoms of leaf rolling and vein banding. The first step for HTS is the extraction of total RNA from leaf samples according to Boom (as described in chapter 2.4). RNA quality and quantity were determined as mentioned (chapter 2.4.1 and 2.4.2). The integrity of RNA extracted (cDNA, nad5) were determined via gel electrophoresis. 60,000 ng RNA was prepared for one HTS sample. The protocol used to digest the remaining DNA was modified from the manufacturer's provided instructions for effective use of the Plant RNA Reagent (Macherey- Nagel Company). All solutions were prepared with sterile RNase-free water and optimized to isolate RNA from approximately 0.1g of plant materials. This was done by rDNAse (Recombinant DNase) digestion which included addition of 60 µl to 90 µl rDNase reaction buffer to extracted RNA samples (30 µl) (ThermoFisher Scientific company) and incubated at 37°C for 10 minutes in agitation at 180-rpm using a Thermomixer (MB-102 Mixing Block, BIOER).

2.12.2 Purification, precipitation and concentration of RNA

The RNA was purified (ThermoFisher Scientific company) by filling up RNA samples to 100 µl with dH$_2$O. A lysis buffer RA1- ethanol (Macherey-Nagel) premix with ratio 1:1 was prepared for each sample of 100 µl RNA. The sample mix was added and it contained 300 µl buffer RNA1 and 300 µl of ethanol (96%-100%). The RNA binding conditions were then adjusted by adding 600 µl of buffer RA1 ethanol-premix to 100 µl RNA sample and carefully mixed by vortexing.

For each preparation, one NucleoSpin RNA Clean-up Column placed in a collection tube, loaded with the lysate (700 µl) and then centrifuged for 30 seconds at 8000 x g (Eppendorf 5810R). The collection tube was then discarded with the column placed in a new tube.

700 µl of buffer RA3 was added to the NucleoSpin RNA Cleaning Column and centrifuged again for 30 seconds at 8000 x g (Eppendorf 5810R). The flow-through was discarded including the DNase and the collection tube was reused again. A second washing was done by adding 350 µl buffer RA3 to the NucleoSpin RNA Clean-up Column and centrifuged for 2 minutes at 8000 x g (Eppendorf 5810R). The NucleoSpin RNA clean-up column was transferred to a nuclease free collection tube (1.5ml, supplied) and the lid was opened for membrane to dry for 3 minutes. The RNA was eluted in 60 µl RNase-free H_2O (re-suspend pellet in dH_2O) and centrifuged at 8000 x g for 1 minute. By this method the DNase was removed.

According to Adams *et al.*, 2009, it is of great advantage in next-generation sequencing and metagenomics analysis to remove the rRNA fraction from the totalRNA before preparation of cDNA library. This was done in the following steps. First the totalRNA was precipitated and concentrated. TotalRNA sample was transferred into a clean RNAse-free 1.5ml micro centrifuge tube. The following components (2.5X sample volumes of 100% ethanol, 1/10th sample (eluted RNA), volume (30 µL for this protocol) of 3M sodium acetate) and 1 µL glycogen (20µg/L) were mixed well at pH of 5.5 and incubated at 80°C for minimum of 30 minutes. Elution of precipitated RNA in DEPC water depended on RNA concentration. The tube containing the totalRNA was then centrifuged again for 15 minutes at 12.000 x g at 4°C and the supernatant was carefully discarded. 500 µl of cold 70% ethanol was added and centrifuged for 5 minutes at 12.000 x g (Eppendorf 5810R) at 4°C and the supernatant was discarded. This was repeated once again and air dried for 5 minutes (SC110 Savant). The totalRNA pellets were then suspended in 10-30 µl DEPC water and placed on ice or stored in -80°C.

2.12.3 Removal of rRNA from totalRNA

The purified pellet was quantified using UV absorbance (NanoDrop One/One[C] micro volume UV/VIS Spectrophotometer (Thermo Fisher Scientific Company) at 260nm. The rRNA species in the totalRNA were depleted by applying the RiboMinus Plant Kits (Thermo Fisher Scientific Company) resulting in a high quality RiboMinusRNA (TotalRNA without rRNA). This kit contains additional plastid components which depletes rRNA from plant species such as wheat, maize, rice, tobacco, tomato as well as birch leaves.

First the magnetic beads with the rRNA directed hybridization probes are prepared. For the preparation of beads in hybridization buffer, RiboMinus magnetic beads were re-suspended in its bottle through short vortexing. 750µl of the bead suspension was pipetted into a sterile, RNase-free 1.5ml micro centrifuge tube. The 1.5 mL micro centrifuge tube were placed ona magnet separator for 1 minute and settled to the tube side that faces the magnet. This was then aspirated and the supernatant discarded. 750 µl DEPC H_2O was added to the beads and re-suspended by slow vortexing followed by placing tube on magnetic separator for 1minute. This step was repeated again and beads were re-suspended in 750µl hybridization buffer.

250 µl beads were pipetted to a new tube at a temperature of 37°C. The remaining 500µl beads were placed on magnetic separator for 1 minute and supernatant gently aspirated and discarded. The beads were re-suspended in 200 µl Hybridization Buffer and kept at 37°C until use. According to RiboMinus Plant Kit the second step in removal of rRNA from the totalRNA was incubation of the tubes containing isolated totalRNA, RiboMinus Probe containing the magnetic beads and hybridization buffer at 70-75°C for 5 minutes for denaturation. Volumes of components are given in Tab. 12.

Tab. 12 Hybridization components for RiboMinus RNA Plant kits

Component	Volume
Total RNA (1-10 µg)	<10 µl
RiboMinus Probe (15pmol/ µL)	10 µl
Hybridization buffer	100 µl
Total	120 µl

Samples were allowed to cool at 37°C slowly over 30 minutes by placing the tube in 37°C Water bath. This was done to promote sequence-specific hybridization. The 120µl was transferred into RiboMinus Magnetic beads (prepared beads) and mix well by pipetting up and down or low speed vortexing.

This was then incubated at 37°C for 15 minutes and during incubation, the contents were gently mixed occasionally and centrifuged after to collect sample at the bottom of the tube. The tube was placed on a magnetic separator for 1 minute to pellet the rRNA-probe complex. The supernatant did contain RiboMinus RNA. 320µl of the supernatant containing the RiboMinus RNA was pipetted to a new tube. The purified RiboMinus RNA was quantitated using UV absorbance at 260nm.

To verify rRNA depletion, agarose gel electrophoresis was done and to verify the efficiency of rRNA depletion in RiboMinus RNA. Agarose gel shows depletion of 18S and 28S rRNA bands as compared to a control sample.

2.12.4 Ds cDNA generation

The last step involves the use of Maxima H Minus Double –stranded cDNA Synthesis Kit (Thermo Fisher Scientific Company) which is the first strand cDNA synthesis. The reaction components were briefly mixed and placed on ice with an appropriate combination of total RNA (Tab. 13).

Tab. 13 Reaction buffer.

Primer	Volume of total RNA
Random hexamer or specific primer (50pmol)	0.5µg-5µg(13µl)
H_2O (nuclease free)	Up to 14µl

For each random hexamer, 13 µl of RiboMinusRNA together with DEPC H_2O was mixed gently and incubated at 65°C for 5 minutes and later placed on ice. A mixture of (4X First Strand reaction mix (5µl) and First Strand enzyme (1 µl)) was mixed, centrifuged and incubated for 10 minutes at 25°C followed by 30 minutes at 50°C. The reaction was then terminated by heating at 85°C for 5 minutes and kept on ice. An immediate second strand synthesis with reaction components is described in Tab. 14.

Tab. 14 Second strand reaction mix.

Reaction components	Volume
First Strand cDNA Synthesis Reaction Mixture	20µl
DEPC H_2O	55µl
5X Second Strand Reaction Mix	20µl
Second Strand Enzyme mix	5µl
Total volume	100µl

A total volume of 100µl reaction buffer was gently mixed, centrifuged briefly and incubated at 16°C in a microbiological incubator (Thermo Fisher Scientific) for 60 minutes. The reaction was terminated and 6µl of 0.5 M EDTA, pH

8.0 was added and mixed gently. This was kept at 4°C until ready to use. To remove RNA residuals from the double stranded cDNA preparation especially if the total RNA was used as a starting material, 10µl of RNase1 was added to the second strand synthesis reaction tube and incubated for 5 minutes at room temperature before proceeding with cDNA purification or stored at -20°C.

20µl of purified cDNA was pipetted into a new tube and sent for HTS to baseclear. At BaseClear an Illumina- ILMN Nextera XT cDNA library was prepared and the sequencing was done by Illumina technology resulting in HiSeq 100 MB paired end data. The company did a "de novo Assembly" (<50 MB) resulting in contigs.

The order of contigs, and the distances between scaffolds were estimated using the insert size information derived from an alignment of the paired-end reads to the draft assembly. Contigs were linked together and placed into scaffolds using SSPACE version 2.3 (Boetzer *et al.*, 2011). Using Illumina reads, gapped regions within scaffolds were (partially) closed using GapFiller version 1.10 (Boetzer and Pirovano, 2012). Finally, assembly errors and the nucleotide disagreements between the Illumina reads and scaffold sequences were corrected using Pilon version 1.21 (Walker, *et al.*, 2014).

2.13 Bioinformatics including genomic database analysis - (NCBI)

2.13.1 De novo assembly and mapping to reference sequences

The HTS contigs of the *Birch leaf-roll associated virus* (BLRaV) were generated by de novo assembly of reads using SPAdes in Geneious prime and baseclear (Köpke, 2019).   The aligned sequences of BLRaV were produced by mapping each of the paired reads from BLRaV to a specific references sequence (BLRaV-Bet- Finland-2014b AR (NC-040635) using Geneious prime (Köpke, 2019).

2.13.2 Database analysis for integrated form of BLRaV

Due to the fact that badnaviruses might integrate into the genome of hosts, the search for additional integrated sequence information especially in DNA databases of birch was essential. A special focus was therefore on the **Whole genome shotgun** (WGS) database for additional sequence information at the RNA. The BLAST program was used to perform sequence similarity searches against a variety of nucleotide or protein queries with nucleotide or protein data bases(Altschul *et al.*, 1990; Ye *et al.*, 2006).

The homepage for BLAST (http://www.ncbi.nlm.nih.gov/BLAST/) consist of nucleotide blast (nucleotides to nucleotides), blastx (which translates nucleotides to proteins), tblastn (which translates proteins to nucleotides), and protein blast (which translates proteins to proteins) (NCBI, 2021). The BLAST server which finds short matches between two sequences and attempts to start alignments from hotspots has diverse set of features which add power to BLAST searching (Altschul *et al.*, 1997).

To gain more sequence information on integration status of the Badnavirus, one out of 11 selected sequences of BLRaV in the general consensus sequence alignment (see Fig. 5) was blasted (BLASTn) against standard database and respective organisms (Appendix 11.3; Table A24) under the program selection *"somewhat similar sequences"*. This was used because *highly similar sequences* (megablast) and *more dissimilar sequences* (discontinuous megablast) did not give any significant results.

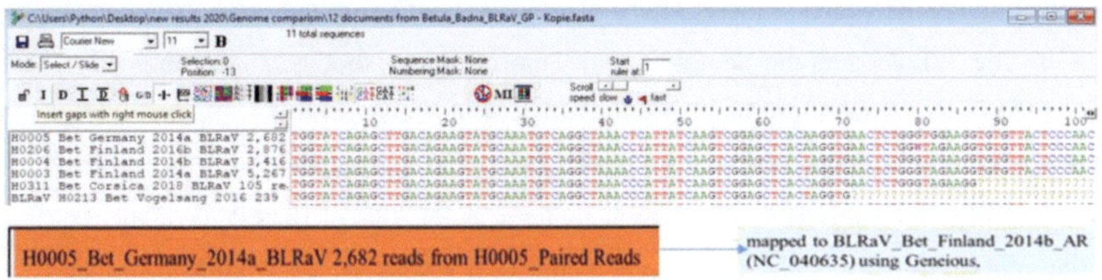

Fig. 5 Consensus sequence of general alignment (partial) of a genome of the *Birch leaf roll associated virus* (BLRaV) (Köpke *et al.*, 2019), mapped to a reference sequence (NC_040635) using Geneious prime. H0005 =number, Bet= *Betula sp.* from Germany in the year 2014, Orange highlighted sequence was investigated using the BLAST tools.

The nucleotide BLAST page provides three selection programs that vary in their sensitivity and speed; highly similar sequence (megablast), more dissimilar sequence (discontiguous megablast), and somewhat similar sequence (blastn). Megablast is intended for comparing a query to closely related sequences and works best if the target percent identity is 95% or more but is very fast. Discontiguous megablast uses an initial seed that ignores some bases (allowing mismatches) and is intended for cross-species comparisons. The Blastn allows a word-size down to seven bases but slow. In this thesis, there was much focus on contig database using the Blast algorithm *"somewhat similar sequence"* (Blastn) because more distantly related sequences information was derived but slower than megablast and discontiguous megablast (NCBI, 2021).With the HTS, raw data was trimmed following a standardized protocol; taking into consideration contigs which is a series of overlapping DNA sequences used to make aphysical map that reconstructs the original DNA sequence of a chromosome or a region of a chromosome. The coverage describes the average number of reads that align to, or "cover," known reference bases. The sequencing coverage level often determines whether variant discovery can be made with a certain degree of confidence at particular base positions.

The score or estimated percentage of genome covered, open reading frames, population level, characterization and contaminant analysis are also factored in HTS analysis. Reads are then mapped to reference genome of viruses or viroids identified using the expected virome after trimming. For unknown viruses, de-novo assembly are used after removal of host sequences and further blasted against the database.

For each of the genomic BLRaV contigs, sequences were copied, blasted via the NCBI and filtered by selecting results which were related to plant viruses affecting *Betula* sp. There was much focus on the Whole genome shotguns and Transcriptome shot gun assembly after filtering since they gave some significant results which could further be investigated. The flow chart (Fig. 6) illustrates the steps used for blasting via NCBI with specific interest on the whole genome contigs and transcriptome shotgun analysis. A BLAST of BLRaV sequence at the nucleotide and protein level has given some significant results which would be further discussed in this thesis.

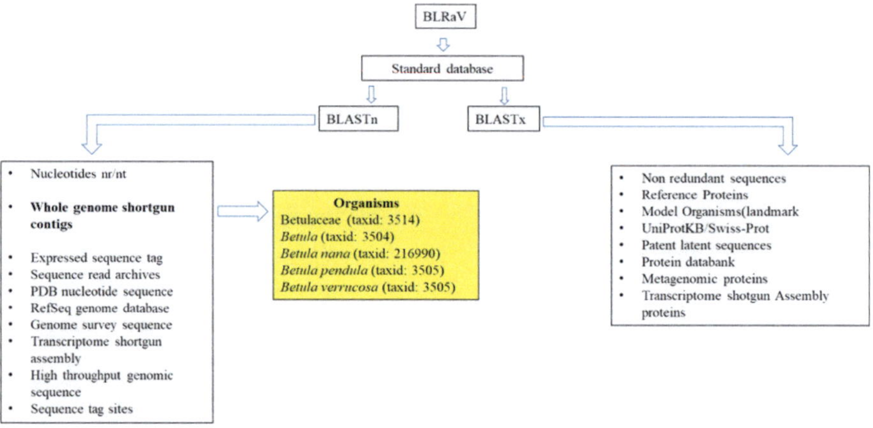

Fig. 6 Flow chart of bioinformatical analysis of the sequence of the genome *Birch leaf roll associated virus* (BLRaV) with accession number (MG686419.1)- Rumbou *et al.*, 2018 with much focus on the **Whole genome shotgun contig** (WGS)-highlighted in yellow

The standard database; nucleotide collection, whole genome sequences (WGS), expressed sequence tags (EST), sequences read archive (SRA), transcriptome shotgun assembly (TSA), High throughput genomic sequences (HTGS), patent sequences (PAT), protein databank (PDB), reference genomic sequences, genomic survey sequences (GSS), sequence tagged sites (STS) were optimized using the parameter somewhat similar sequences.

38

2.13.3 Nucleotide collection (nr/nt)

The nucleotide collection is non redundant which has been merged into one entry and consists of (GenBank, European molecular Laboratory(EMBL), DNA Data Bank of Japan (DDBJ), Protein databank (PDB), RefSeq) sequences, but excludes (expressed tag sequence(EST), sequence tagged sites (STS), genomic survey sequences (GSS), whole genome sequences (WGS), transcriptome shotgun assembly (TSA) patent sequences as well as phase 0, 1, and 2 high throughput genomic sequences (HTGS), longer than 100Mb (NCBI, 2021).

2.13.4 Whole genome sequences (WGS)

Using the NCBI as a benchmark, the BLRaV sequence was blasted on the nucleotide level under whole genome sequence against organisms such as; *Betulaceae* (taxid: 3514), *Betula* (taxid: 3504), *Betula nana* (taxid: 216990), *Betula pendula* (taxid: 3505), *Betula verrucosa* (taxid: 3505) which were optimised using *somewhat similar* program.

There was much focus on the whole genome shotgun contigs sequence (WGS), because it provides information on genome fragments of especially plants and was expected to give information of integrated viruses such as the Badnavirus and Caulimovirus on the DNA level. The results obtained from Blasting (WGS) against organisms were filtered and significant accession numbers were selected. Each of the accession numbers were converted to FASTA and Graphics. Using the graphics tool provided by NCBI, each accession number was configured into tracks. Search for integrated viruses were done by moving forward and reverse checking ORFs. For every long ORF, a target range was selected and blasted against plants and viruses. An example of a flowchart with different accession number is shown (Fig. 7).

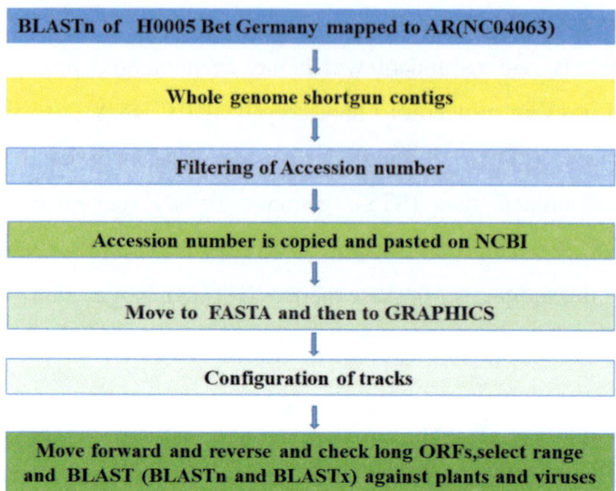

Fig. 7 Flowchart of bioinformatical analysis of the genome *Birch leaf roll associated virus* (BLRaV) sequence with accession number (MG686419.1), Rumbou *et al.*, 2018 with much focus on the Whole genome shotgun contig (WGS).

3 Results

3.1 Symptomatology in birches of selected stands

Virus-like symptoms in birches from Berlin, Wildberg/Nagold and Rovaniemi were categorized into groups of chlorosis, chlorotic spot, ring spot, intercostal chlorosis, leaf mosaic, line pattern, mottling, necrosis, oak leaf pattern, necrosis, ringspot, leaf variegation, vein banding, leaf deformation and leaf roll. Due to the complexity of birch leaf symptoms, a general description has been shown in (Appendix 11.4; Fig. A1)

3.1.1 *Betula* leaf symptoms -Berlin -Germany

Most *Betula* stands from Berlin affected by viruses showed symptoms of chlorosis, vein banding, and necrosis which seems to be the final stage of chlorosis, especially if vein banding was involved. Symptoms at different developmental stages were observed during visual rating and this made interpretation of leaf patterns difficult. Leaf symptoms (72 *Betula* sp.) from Berlin were rated based on various categories described in (Appendix 11.4; Fig. A1). A graph shows intercostal chlorosis being prominent with leaf mosaic and line pattern recording the least (Fig. 8).

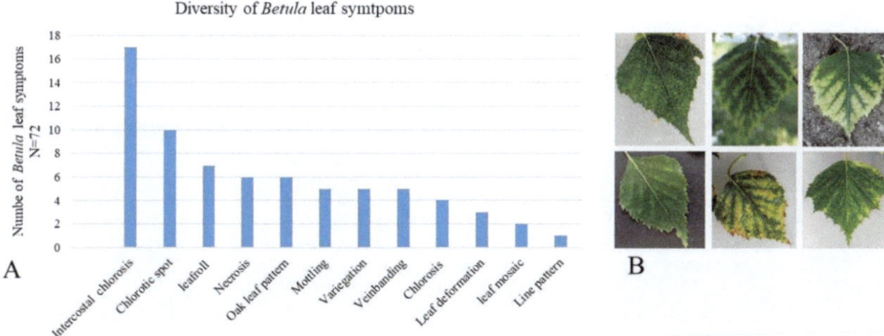

Fig. 8 (A) Diversity of leaf symptoms ratings of *Betula* stands in Berlin, 2017. Number of *Betula* leaf symptoms (N= 72). (B) Examples of prominent symptoms of intercostal chlorosis observed.

Different forms of chlorosis such as ringspot (A- E56736), oak leaf pattern (B- E54088), leaf roll (C-E54047), vein banding (D-54065), mottling with ringspot (E- E54085), chlorotic spot (F- E54030), vein chlorosis (G- E54072), and vein banding (H- E54056) were observed in different Berlin stands (Fig. 9).

Fig. 9 Selection of prominent symptoms observed during tree rating in May, 2017 in Berlin (Germany). Most of the tree samples were collected at Parks. Obvious viral symptoms as well weak symptoms were observed. Symptom descriptions, location, sample number are given: (A) Ring spots, Neukölln, E56736; (B) Oak leaf pattern, Reinickendorf, and E54088; (C) Leaf rolling and chlorosis, Köpenick, E54047; (D) Vein banding, Pankow, E54065; (E) Mottle and ring spots, Reinickendorf, E54085; (F) Chlorotic spots, Charlottenburg-Wilmersdorf, E54030; (G) Vein chlorosis, Mitte, E54072; (H) Vein banding, Köpenick, E54056.

3.1.2 Leaf symptoms from *Betula* stands -Wildberg/Nagold -Germany

Out of 117 *Betula* trees in the Wildberg/Nagold seed plantation, 72 trees were selected according to requirements of the Pollen PALS project (DFG BU 890/31-1 (655850). Two (2) additional trees (E58128 and E58145) had symptoms of vein banding, intercostal chlorosis and leaf roll (Fig.10). There were very diverse insect populations compared to Berlin environment. Due to higher insect populations, there were insect bites on leaves masking viral symptoms in the leaves during vegetation period.

Fig. 10 Impressions of suspected viral *Betula* leaf symptoms observed in Wilberg/Nagold seed plantation Baden-Württemberg (Germany) June, 2019. E58125 insect infestation with corresponding insect bite (1), E58145 vein banding (2), E54120 intercostal chlorosis (3), E58125, leaf roll and vein banding (4), chlorotic spots and insect bites (5 and 6), E58125, leaf deformation (7 and 8) (Division of Phytomedicine -HUB, 2019).

3.1.3 HTS in correlation to *Betula* leaf symptoms -Wildberg/Nagold-Germany

Two *Betula* tree samples (E58128 and E58145) with symptoms of chlorosis, vein banding and leaf roll (see Fig. 10) were associated with BLRaV. Pooled samples of two symptomatic leaves from birch trees (E58128 and E58145) selected were used for total RNA isolation and sent for High-throughput sequencing (HTS). The resulting cleaned reads were then assembled into contigs, generating between 739, 7220 and 7,226 respectively. Contigs longer than the 300 nucleotide cut-off were then annotated by BLASTN and BLASTX against the GenBank database.When needed, contigs representing the same genus badnavirus were manually assembled into larger scaffolds and scaffolds polished by re-mapping of reads on the scaffolds using CLC Genomics Workbench as reported by Rumbou *et al.*, 2018.

HTS results from the pooled samples (E58128 and E58145) produced significant hits of plant viruses with E-values of zero and could be closely related to BLRaV with accession number MG686421.1. Different E-values, Query (Contig 40, Contig 41 and Contig 5978), hosts, viruses (taxid: 10239), sequence length, amount of contigs (how many viruses were addressed to the taxid result) and individual reference viruses are shown (Tab. 15) after BLASTx.The Contig 5978 resulted in small basepairs lenght and could be a new possible Badnavirus but was not further investigated in this study.

Tab. 15 HTS results from two different *Betula* sp. (E58128 and E58145) with specific symptoms of (chlorosis, vein banding and leaf roll) were closely associated with BLRaV. Expected value of hit, query, host, sequence length, contig amount with BLASTx on viruses (taxid: 10239) and individual BLASTx. Grey colour represents plant viruses. Viruses in (yellow colour) were found in the HTS but were not relevant in the investigation because they belong to vertebrae, invertebrate and fungi.

E-Value	Query	Host	Blastx database nr: Viruses (taxid:10239)	Sequence Length in basepairs	Contig amount	Blastx database nr: individual
0	Contig 40	plants	Viruses; Ortervirales; Caulimoviridae; Badnavirus	7226	1	Badnavirus (*Birch leaf roll-associated virus*)
0	Contig 41	plants	Viruses; Ortervirales; Caulimoviridae; Badnavirus	7220	1	Badnavirus (*Birch leaf roll-associated virus*)
4,89E-51	Contig 5978	plants	Viruses; Ortervirales; Caulimoviridae; Badnavirus	739	1	Badnavirus
1,2E-16	Contig 766	fungi| plants| invertebrates	Metaviridae; Metavirus	2462	32	plant: Transposon Tf2-12 polyprotein [Vitis vinifera]
0,000646	Contig 81	invertebrates	Metaviridae; Errantivirus	5664	14	plant: *Betula platyphylla* salt induced long non-coding RNA 3 lncRNA
1,18E-07	Contig 95	vertebrates	Retroviridae; Spumaretrovirinae; Equispumavirus	5374	47	plant: *Actinidia chinensis* DNA, Y-specific genomic marker fifth one
6,96E-28	Contig 2166	fungi	Viruses; Riboviria; Narnaviridae; Mitovirus; unclassified Mitovirus	1470	3	fungi virus: RNA-dependent RNA polymerase [*Ambrosia artemisiifolia mitovirus 1*]

Tab. 16 HTS results from two different *Betula* sp. with specific symptoms of (chlorosis, vein banding and leaf roll) associated with BLRaV (E58128 and E58145). Percentage identity matrix of 92% and 94% for the H0515_contig41 BLRAV and H0515_contig40 BLRaV was significantly related to BLRaV at the nucleotide level with the accession number (MG686421.1) *Birch leaf roll-associated virus* isolate BpenGer407526, complete genome. The percentage identity matrix were calculated using the software Bioedit alignment after trimming of nucleotides.

Seq->	MG686421.1 Birch leaf roll-associated virus isolate BpenGer407526, complete genome	H0515_contig41 BLRAV	H0515_contig40 BLRaV
MG686421.1 Birch leaf roll-associated virus isolate BpenGer407526, complete genome	ID	92%	94%
H0515_contig41 BLRAV	92%	ID	93%
H0515_contig40 BLRaV	94%	93%	ID

3.1.4 *Betula* leaf symptoms -Rovaniemi -Finland

Leaf symptoms observed in Rovaniemi were comparable to Berlin. Samples with symptoms of leaf roll, intercostal chlorosis, ringspots and necrosis dominated in all areas of sample collection. They were prominent in most trees. Based on the 26 *Betula* sp. sampled leaves, 42% of the investigated leaves had symptoms of intercostal chlorosis with the least rating being chlorotic spot. Necrosis recorded 11%. There was rating of asymptomatic leaves recording 8%.

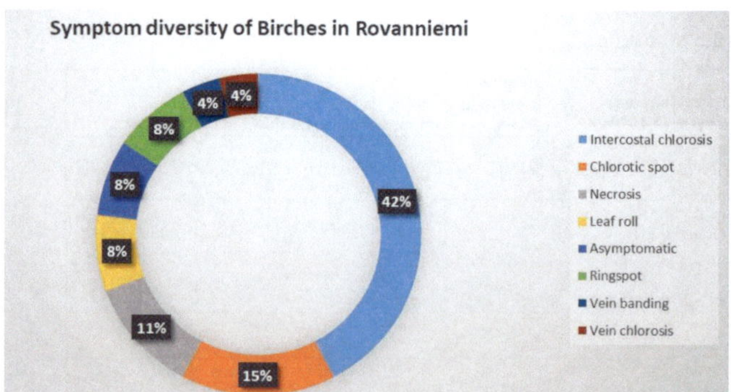

Fig. 11 *Betula* sp. leaf symptoms ratings (intercostal chlorosis, chlorotic spot, necrosis, leaf roll, ringspot, vein banding, vein chlorosis and asymptomatic leaves) ratings from Rovaniemi -Finland in June, 2017.

44

3.2 Results of RNA extraction and RT-PCR

3.2.1 Separation of RNA by gel electrophoresis

The RNA from symptomatic *Betula* leaf material from (Berlin, Rovaniemi and Wildberg-Nagold) recorded an average concentration of 300ng/ul. There was presence of three and more distinct bands visible under the UV-light which indicates a high quality of RNA after running agarose gel for 45 minutes at 70volt (Fig. 12).

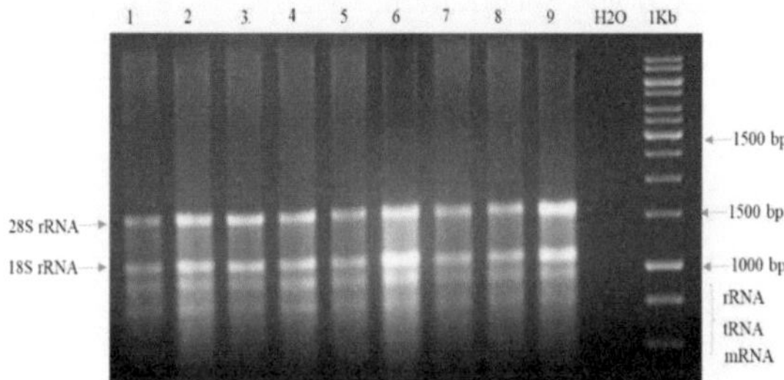

Fig. 12 An example of agarose gel electrophoresis showing total RNA extracted from *Betula* leaves with different E numbers (1=E54030, 2=E54032, 3=E54036, 4=E54038, 5=E54040, 6=E54042, 7=E54055, 8=E54065, 9=E54095, H_2O =negative control) using 1.0% agarose gel and SERVA DNA Stain G observed under UV-light. Marker: 1Kb DNA ladder. E-numbers for all leaf samples are shown in Appendix 11.3; Tab. A20, Tab. A21 and Tab. A22.

The top band represent 28S ribosomal RNA (rRNA) which normally runs at ~4.8 kb; the middle band represent 18S rRNA at ~2.0 kb; and the third band represents 5.8S (154 nt) and 5S (117 nt) RNA. The third band indicates tRNAs and smaller rRNAs (5S, 5.8S and 4.5S rRNA). The appearance of the 28S and 18S had comparable signal intensity. There were no smeared bands or a 28S:18S band intensity ratio below 2:1 indicating no degradation of RNA. Fig. 12 shows a high-quality RNA extraction. The size of different RNA types was estimated by doing a comparison using 1Kb genomic DNA as the reference marker.

3.2.2 Results of RT-PCR detecting the nad5 transcript in gel electrophoresis

To rule out false negative results, one primer pair specific to plant mitochondrial NADH dehydrogenase (nad5) gene was applied to amplify the product of plant nad5 mRNA (Menzel *et al.*, 2002). Due to the varying quantity of RNA isolation (Appendix 11.3; Tab. A20, Tab. A21 and Tab. A22), 1 µl to 5 µl RNA was used for the cDNA synthesis. Examples of nad5 PCR product resulting from cDNA used as a template DNA of investigated samples after gel electrophoresis with a no template control (H$_2$O) and 50bp reference marker at 181bp is shown in (Fig. 13).

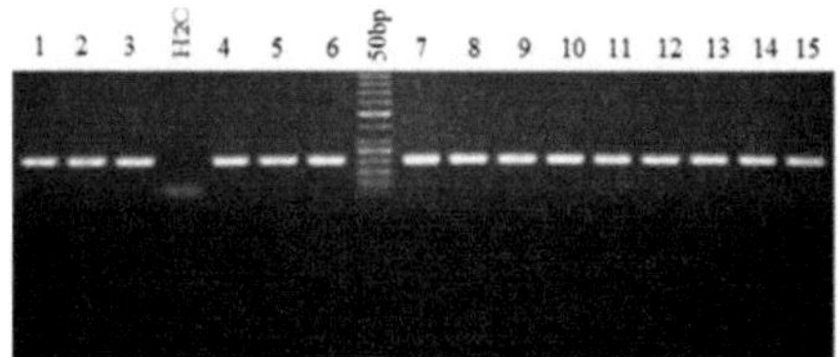

Fig. 13 (1.0%) Agarose gel electrophoresis of PCR product amplified with primers for the nad5 gene transcript.1=E54030, 2=E54032, 3=E54034, 4=E54036, 5=E54038, 6=E54039, 7=E54050, 8=E54052, 9=E54055,10=E54060,11=E54062, 12= E54065, 13=E54070, 14=E54090,15=E54095 using pMMLV-Rtase (Phytomedizin) and random hexamers in reverse transcription, (H$_2$O) no template control (NTC) and 50bp marker. E- numbers are shown in Appendix 11.3; Tab. A20, Tab. A21 and Tab. A22.

3.2.3 Result of RT-PCR - BLRaV

RT-PCR results for BLRaV with primer targeting CP (BadnaK forward/reverse) at an expected size of 300bp (see Fig. 14). PCR products were allowed to run for 45 minutes using 1.5% agarose gel at 70V and intercalating DNA star dye observed under UV light.

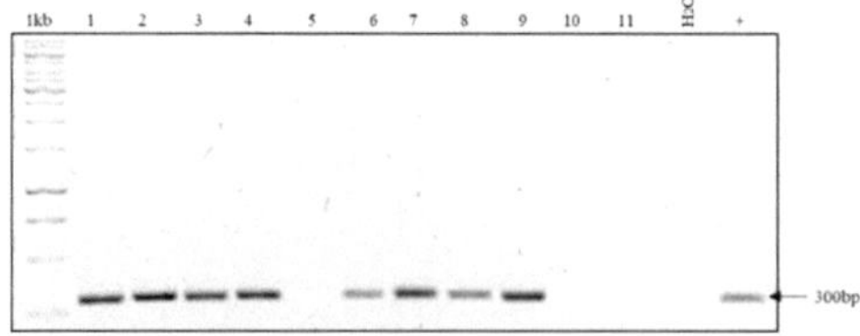

Fig. 14 Agarose gel electrophoresis (1.5%) of BLRaV from Berlin samples after RT-PCR; (1Kb= reference marker, 1=E54035, 2=E54039, 3=E54042, 4=E54045, 5=E54048, 6=E54050, 7=E54052, 8=E54065,9=E54067, 10=E54089, 11= E54090, H20= (NTC), (+) =E54036. For (5=E54048, 10=E54089 and 11= E54090), infection could not be

detected.

The abundance and distribution of BLRaV in birch samples from Berlin stands,Wildberg/Nagold and Rovaniemi stands are summerized in (Appendix 11.3, Tab. A20, Tab. A21 and Tab. A22). BLRaV was confirmed by RT-PCR in only trees (E58128 and E58145) but were not present in other trees of the seed plantation. These BLRaV distribution indicates how urban trees on streets and Parks of Berlin are affected, hence a need for Urban management to consider viral screening of young tree seedling for viruses.

3.2.4 BLRaV and CLRV detection in *C. quinoa*

A mechanical transmission was undertaken to transfer BLRaV from infected birch leaves (E54094- TT SEAL 0407510 and E54095- TT SEAL 0407519) to indicator plant (*C. quinoa*). After testing all indicator plants using protocol as described in chapter 2.7.1 to 2.7.4 for CLRV, BiCV, ApMV and BLRaV by RT-PCR respectively, a single infection of BLRaV was detected by RT-PCR in *C. quinoa* (3=E54095) with a primer combination of BadnaSG (Forward/Reverse) at an amplicon size of 242bp (see Fig. 15A). A second *Chenopodium* plant (1=E54094) had mixed infection of CLRV and BLRaV by RT- PCR (Fig. 15).

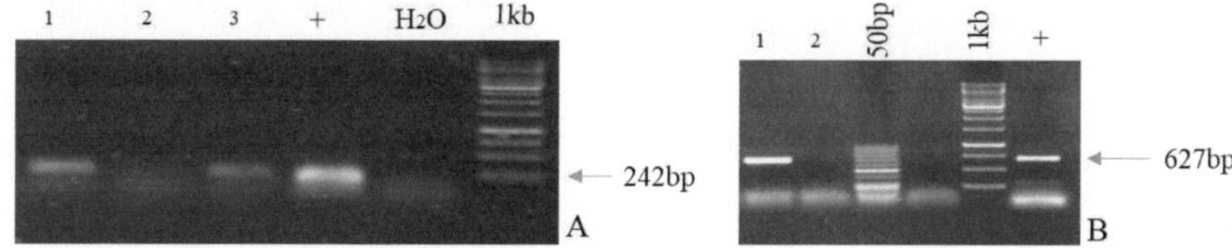

Fig. 15 BLRaV infection was confirmed in bioassay (*C. quinoa*) using RT-PCR. (A) represent gel electrophoresis of BLRaV detection by RT-PCR at 242bp at 70V for 45 minutes using 1.0% agarose gel; 1=E54094, 2=E54080 and 3=E54095 represents bioassay plants, (+) represent positive control from E54036. (B) Represent a gel electrophoresis of CLRV (RT-PCR) with a primer combination of (CLRV CP 350F/CP350R) at 627bp using 1.5% agarose gel; 1= E54094 and 2= E54095 represents bioassay plants, (+) represents positive control from E54030, dH$_2$O=(NTC), 50 bp;1Kb reference marker.

Bioassays confirmed *C. quinoa* to be a suitable test plant for BLRaV infection developing distinct symptoms like ringspots and intercostal chlorosis (Tab. 17). Other indicator plants (*Nicotiana benthamiana, Nicotiana tabacum , var Samsun*) showed no symptoms and BLRaV was not detected.

BLRaV and CLRV were first detected in systemic leaves of *C. quinoa* which developed chlorotic symptoms. However, symptoms cannot be attributed solely to BLRaV and CLRV as the herbaceous hosts were not tested for new viruses such as (Birch caulimovirus-like sequence, Birch Idaeovirus, Birch benyvirus-like sequence,) that may have been present in birch tree. *C. quinoa* which tested BLRaV positive was transferred to new *C. quinoa* which resulted in a positive result. BLRaV was also detected in *C. amaranticolor* after first transfer. Both first and second transfer using *N. bethamiana* was however not successful.

Tab. 17 Bioassay experiment results from first and second transfer. Symptomatic birch leaf samples (E54094, E54095 and E54080) were transferred to different bioassay plants (*C. quinoa, C. amaranticolor and N. benthamiana*).

E-numbers	Indicator plants	Symptoms on indicator plants	First transfer from *Betula* leaf sample to bioassay plants		Second transfer from bioassay plant to a new bioassay plants	
			BLRaV PCR detection	CLRV PCR detection	BLRaV PCR detection	CLRV PCR detection
E54094	*Chenopodium quinoa*	Intercostal chlorosis	+	+	+	+
	C. amaranticolor	Intercostal chlorosis	+	+	+	+
	Nicotiana benthamiana	No symptoms	-	-	-	-
E54095	*C. quinoa*	Intercostal chlorosis	+	-	+	-
	C. amaranticolor	Intercostal chlorosis	+	-	+	-
E54080	*C. quinoa*	No symptoms	-	-	-	-
	N. benthamiana	No symptoms				

3.2.5 Result of RT-PCR - (CLRV)

Betula leaf samples from (Berlin, Nagold and Rovaniemi) were tested with a primer combination of CLRV-CP350F/CLRV-CP977R and CLRV-CP188F/CLRV-CP350R targeting partial sequences of CLRV-CP.

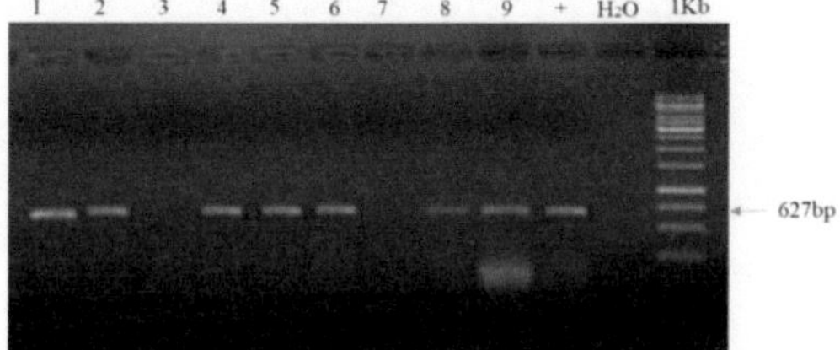

Fig. 16 An example of agarose gel electrophoresis for CLRV detection from Berlin samples by RT-PCR; 1=E54030, 2=E54031, 3=E54032, 4=E54034, 5=, E54059 6=E54060, 7= E54061, 8= E54067, 9=E54072 showed virus-specific amplification of the expected size of 627bp after 45 minutes at 70V. H₂O=NTC, (+) = E54031 using (1%) agarose gel. 3=E54032 and 7=E54061 CLRV were not detected. RT-PCR for all birch leaf samples investigated for CLRV is listed in (Appendix 11.3, Table A20, Table A21 and Table A22).

3.2.6 Result of (BiCV) RT-PCR

With an optimized annealing temperature of 53°C, a primer combination of CarlaK Forward/Reverse gave an amplification of partial coat protein of BiCV at 197 bp after gel electrophoresis for 45 minutes at 70V.

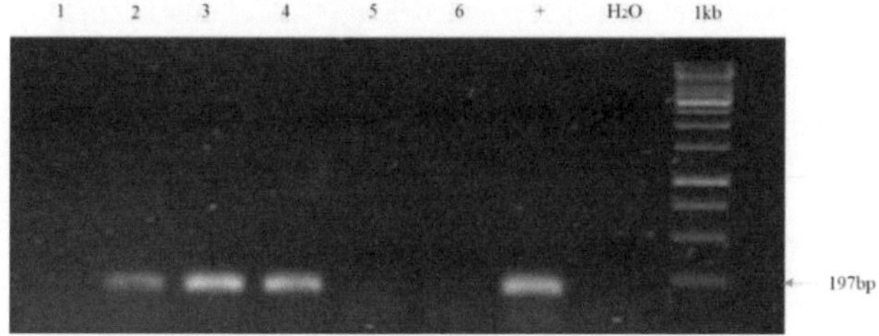

Fig. 17 An example of agarose gel electrophoresis of BiCV after RT-PCR; 1=E54030, 2=E54034, 3=E54047, 4=E54088, 5= E54045, 6= E54046 were amplified at the expected size of 197bp. H₂O = NTC, (+) = E54035 and 1Kb marker after 45 minutes at 70V using (1%) agarose gel. In E54030 and E54045, E54046, BiCV could not be detected. (Appendix 11.3, Table A20, Table A21 and Table A22) shows results on all samples investigated for BiCV. BiCV were not detected in Wildberg/Nagold samples.

3.2.7 Result of (ApMV) RT-PCR

With a well-designed primer and PCR protocol, PCR fragments of ApMV were detected at an expected size of 204bp after running gel electrophoresis for 45 minutes at 80V using (1%) agarose gel.

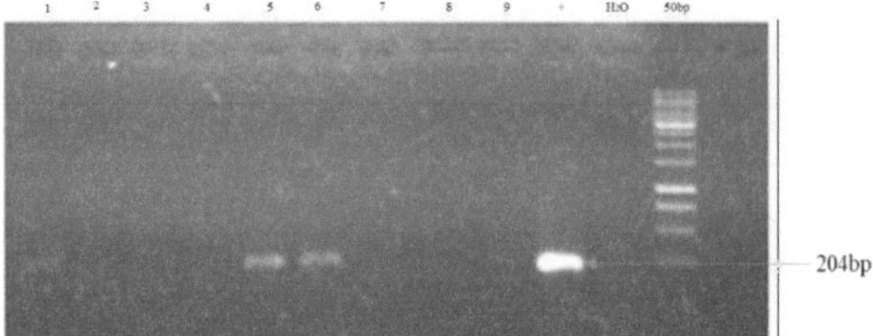

Fig. 18 An example of agarose gel electrophoresis of ApMV after RT-PCR for samples from Berlin; Samples (1=E54039, 5=E54052 and 6=E54086) were detected at the expected size of 204bp. H_2O= (NTC) and (+) =E54033 after 45. No ApMV were detected in Wildberg/Nagold samples. Results on all samples investigated for ApMV is listed (Appendix 11.3, Tab. A20, Tab. A21 and Tab. A22).

3.2.8 Gel electrophoresis of PCR for partial Benyvirus-like sequence

The sequence information from High through-put sequencing (HTS) of *Betula* sp. has given hints of unknown data of potential partial viral genomes. PCR primers were designed based on this information for the detection of viral signatures with similarity to reference sequences from NCBI database. The confirmation of PCR products by Sanger sequencing was provided by Division of Phytomedicine-HUB. Two contigs (01516 and 01856) with similarity to benyviruses were found in HTS samples from 2015 (Finland in tree Bpub20_15396) and provided by Division of Phytomedicine-HUB. For RNA 2 to RNA5 no information could be extracted from HTS data so far.

A first set of primers was designed based on the two contigs originating from RNA 1. The primers named Beny 1856 Forward and Beny 1856Reverse were validated in this study at the amplicon size of 254bp by gel electrophoresis. For the primers from contig01516 the Sanger sequencing was not successfully. The RNA1 sequence was bioinformatically extracted from the HTS dataset by Division of phytomedicine-HUB. Work on confirmation of this artificial sequence is still in progress and therefore the need to further adjust the PCR protocols in order to achieve a specific result. Reference sequences from Benyvirus group were aligned to identify conserved regions for a primer design for targeting the genus level.

The conserved area was found in replicase regions (BenyRNA1Rep5534 FW and BenyRNA1Rep5941 R). Agarose gel (1.5%) electrophoresis gave an unspecific amplification product of partial Benyvirus- like signatures amplified from birch RNA as confirmed by Sanger sequencing of PCR product. Cloning of the unspecific product was not done. The expected fragment 254bp included different amplification products. RT-PCR was not optimized (Fig. 19). The appeared fragment on gel electrophoresis could not confirm the amplification of replicase gene by Sanger sequencing.

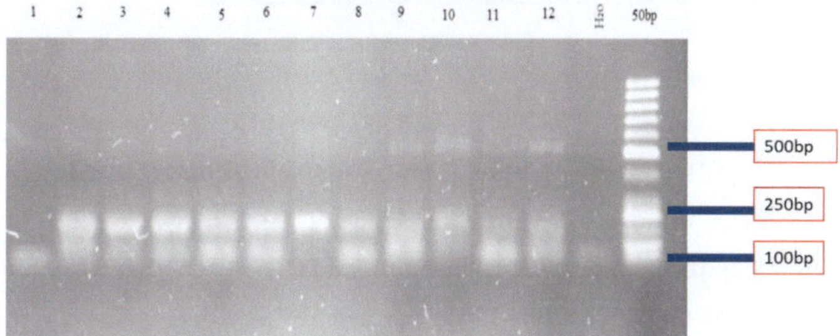

Fig. 19 An example of agarose gel electrophoresis of unspecific partial hit of Birch benyvirus-like signature after RT-PCR; Using a (primer combination of Beny1856F/R) after 45 minutes at 80V using 1.5% agarose gel, (1=E58124, 2=E58125, 3=E58126, 4= E58127, 5=E58129,6=E58130,7=E58131,8=E58132,9=E=E58133,10=E58134,11=E58135,12=E5813 6) were amplified but was not the expected size of 254bp. H₂O=NTC. Results on all samples for birch Benyvirus-like sequence detection listed in (Appendix 11.3; Tab. A22).

The optimization of Benyvirus-like PCR did not give a specific amplification product. The RNA1 was artificially assembled by HUB-Phytomedicine. For birch benyvirus-like sequence contig1856, confirmation by Sanger sequencing wasn't done because contig1516 was confirmed by RT-PCR and Sanger Sequencing. The contig1856 coverage was significant in the HTS dataset. Based on the HTS data the RNA1 from the birch benyvirus-like sequences was assembled in this study.

3.2.9 Birch Capillovirus-like sequence -RT- PCR

HTS data provided contigs with certain similarities to capilloviruses. With an expected size of 469bp, a partial amplification of Capillovirus-like signature was determined by RT-PCR. Capillovirus-like sequences were always detected in most *Betula* samples from Berlin but not always from Wildberg/Nagold samples (Fig. 20). For the first time, 5 out 10 *Betula* trees, Capillovirus-like sequence could not be confirmed even though the amplification of the nad5 transcript showed no inhibition PCR.

They were confirmed by Sanger sequencing by Division of Phytomedicine-HUB (Appendix 11.4; Figure A2)

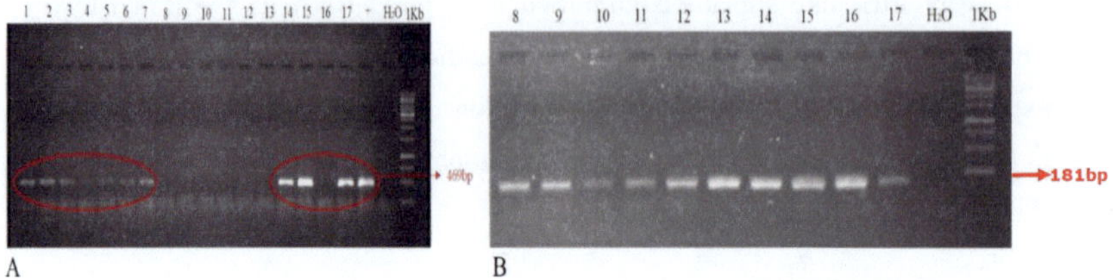

Fig. 20 (A) An example of agarose gel electrophoresis of Capillovirus-like sequence after RT-PCR; 1=E58126, 2=E58127, 3=E58128, 4= E58129, 5= E58130, 6=E58131, 7= E58132, 14= E58139, 15=E58140 and 17=E58142 were amplified from coat protein at the expected size of 469bp after 45 minutes at 80V using 1.5% agarose gel. H_2O= NTC, 1Kb marker and (+) =E58088, (B) nad5transcript amplification resulted in an amplified PCR product (Sample 8=58133, 9=58134, 10=58135, 11=58136, 12=58137, 13=58138 and 16= E58141) without showing inhibition. Results on all samples for birch Capillovirus-like sequence detection by RT-PCR is listed in (Appendix 11.3; Tab. A22).

3.2.10 Birch Caulimovirus-like sequence -RT- PCR

A PCR protocol with an annealing temperature of 53°C showed an amplification of birch Caulimovirus-like sequence from the partial RNA1 polyprotein with a specific size of 162 bp. Detection of these viral signatures by RT-PCR was common for Berlin and Wildberg/Nagold samples. PCR products were specific and could be confirmed by Sanger sequencing by Phytomedicine- HUB (Appendix 11.4; Figure A4 and Figure A5).

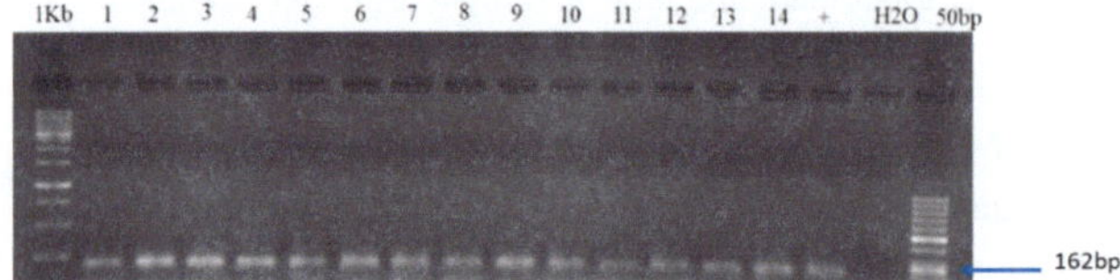

Fig. 21 An example of gel electrophoresis of specific partial hit of Birch Caulimovirus-like by RT-PCR; 1=E58088, 2=E58089, 3=E58090, 4= E58091, 5= E58092, 6= E58093, 7= E58094, 8=E58095, 9= E58096, 10=E58097, 11=E58098, 12=E58099, 13= E58100, 14=E580101 were amplified at the expected size of 162bp after 45 minutes at 80V using 1.5% agarose gel. H_2O=NTC, marker (1Kb and 50bp) and (+) =E58099. Results on all samples investigated for birch Caulimovirus-like sequence is listed in (Appendix 11.3; Tab. A22).

3.2.11 Birch Deltapartitivirus-like sequence -RT- PCR

For Deltapartitivirus-like sequences, PCR product of expected size 324bp was detected but Sanger sequencing could not confirm Deltapartitivirus in the PCR product from Wildberg/Nagold. Partial amplification of Deltapartitivirus-like sequence with a specific size 324bp of the RNA1 was detected by RT-PCR (Fig. 22) using 1.5% agarose gel at 80V for 45minutes. Specific fragments were amplified in 2018 by Phytomedicine-HUB (E56736 from Berlin Rudow) and was confirmed by Sanger sequencing (Appendix 11.4; Fig. A6). Even though the PCR product showed the expected size of 324bp, an approach to sequence PCR products from Wildberg/Nagold couldn't confirm the existence of the Deltapartitivirus. Therefore the need to further adjust PCR protocols and repeated.

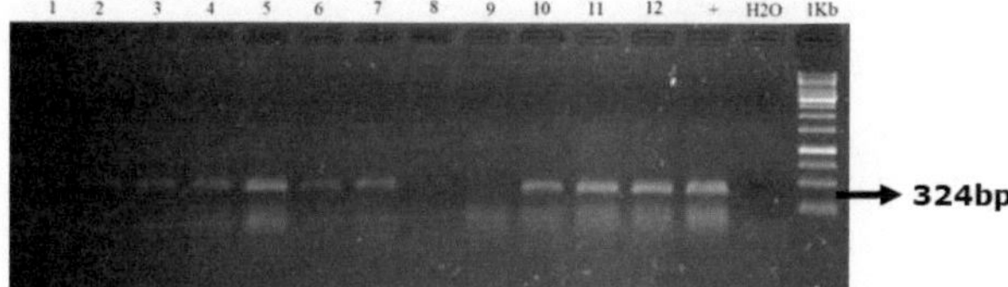

Fig. 22 An example of (1.5%) agarose gel electrophoresis of specific partial amplification of Birch Deltapartitivirus after RT-PCR; 1=E58088, 2=E58089, 3=E58090, 4= E58091, 5= E58092, 6= E58093, 7= E58094, 8=E58095, 9= E58096, 10=E58097, 11=E58098, 12=E58099, 13=E58100, 14=E580101 were amplified at the expected size of 324 bp of the RNA1. H_2O=NTC, (marker; 1Kb) and (+) =E58092.

3.3 Detection of plant viruses infecting *Betula* sp. -Berlin

The general summary of the four detected viruses (CLRV, BLRaV, BiCV, ApMV) infecting *Betula* sp. stands on streets and parks from eight Berlin districts by RT-PCR is shown in Fig. 23.

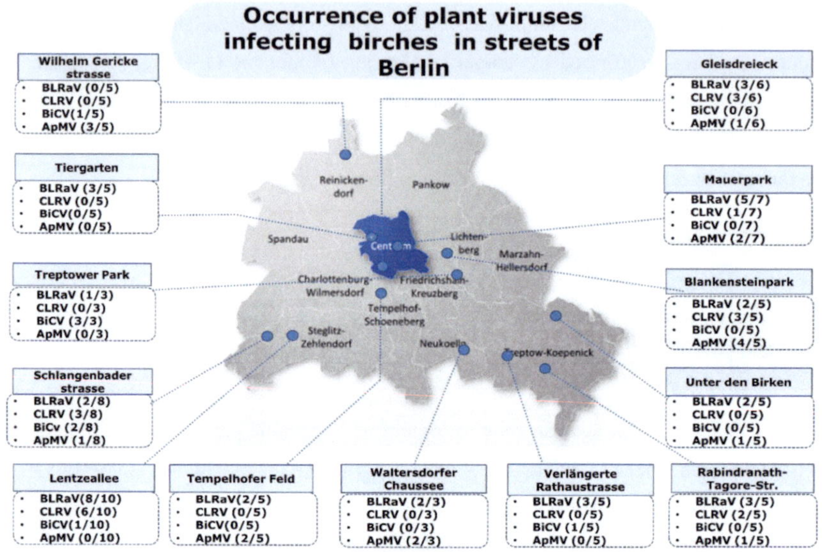

Fig. 23 Virological examination of diseased and degenerating birches in eight districts of Berlin, May- June 2017. At least one of the four known viruses (BLRaV, ApMV, CLRV and BiCV) was detected in *Betula* stands by RT-PCR. The number on the left represents rate of infection of each virus. The numbers on the right represents number of investigated trees. BLRaV were detected in all *Betula* stands with the exception of Wilhelm Gericke Strasse. (Grünflächenamt, Berlin, 2016).

3.3.1 Single and mixed viral infection of *Betula* sp. stands -Berlin

Out of the four tested plant viruses, BLRaV recorded with 55.6% (40) as the most prominent virus in the investigated trees followed by CLRV with 26.4% (19), ApMV with 22.2% (16) and BiCV 8.3% (6). Infection by two different viral species was recorded for BLRaV/CLRV 12.5% (9), BLRaV / ApMV 11.1% (8), CLRV/ApMV 2.8% (2), BLRaV /BiCV 1.4% (1), and CLRV/BiCV1.4% (1). The following combinations of viruses were found in two single trees: BLRaV/BiCV/CLRV and BLRaV/CLRV/ApMV. There was no specific detection of the four investigated viruses in 14 (19.4%) symptomatic birches.

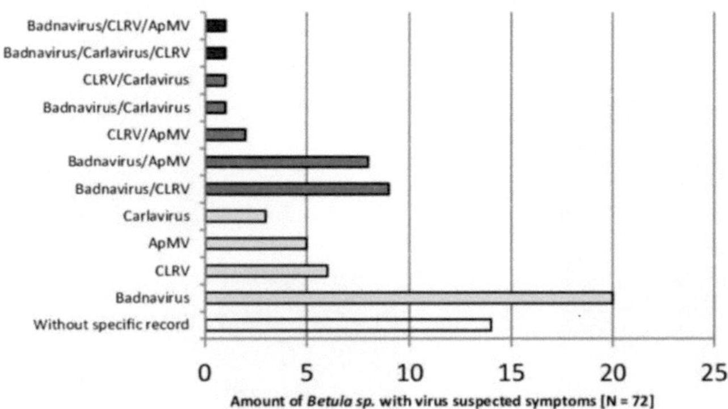

Fig. 24 Graph showing single and mixed viral infection in symptomatic leaves of *Betula sp.* in Berlin detected using RT-PCR. Trees without specific record, Single infection; two viral species infection and three viral species infecting birches were recorded (Opoku *et al.*, 2018).

3.4 Detection of plant viruses infecting *Betula* sp -Wildberg/Nagold

Out of the 71 *Betula* leaf samples, BLRaV was detected in two (2) trees. The remaining leaf samples had none of the four prominent viruses (BLRaV, BiCV, ApMV and CLRV). There was however an amplification product of RT-PCR with primers designed in this study for so far unknown plant viral signatures (Benyvirus-like signatures, Birch Caulimovirus-like signature, Capillovirus- like signature and Deltapartitivirus signature (Fig. 25). Designed primers (from this study) were used to screen birch leaf samples from Wildberg/Nagold seed plantation to ascertain if they would exhibit some traits of these new potential viruses and to confirm the true nature of the sequences in the original birch trees. Sanger sequencing of PCR products amplified for the Benyvirus-like signatures and Deltapartitiviurs signature in trees from Wildberg-Nagold was not successful but was succesful for Birch capillovirus and Birch caulimovirus signatures.

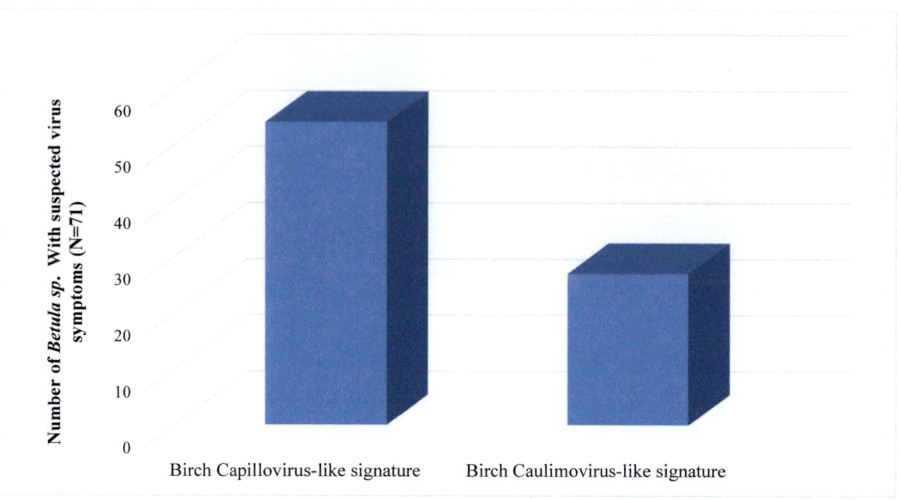

Fig. 25 Amplification product of partial hit of plant viruses from *Betula sp.* in Wildberg-Nagold seed plantation. Number of *Betula* species investigated (N=71). Only the Capillovirus-like signature and the Caulimovirus-like signature were confirmed by Sanger sequencing for leaf samples fromWildberg-Nagold.

3.4 Detection of plant viruses infecting *Betula* sp -Rovaniemi -Finland

Detection of plant viruses affecting birches in Rovaniemi (Finland) had BLRaV dominating in areas of sample collection. Out of the (26) *Betula* sp. investigated, single infection by BLRaV were detected in (23) trees. CLRV were detected in (2) birch samples whiles ApMV detected in (1) birch sample. No BiCV was detected (Fig. 26). There was mixed infection of BLRaV /CLRV and BLRaV/ApMV respectively.

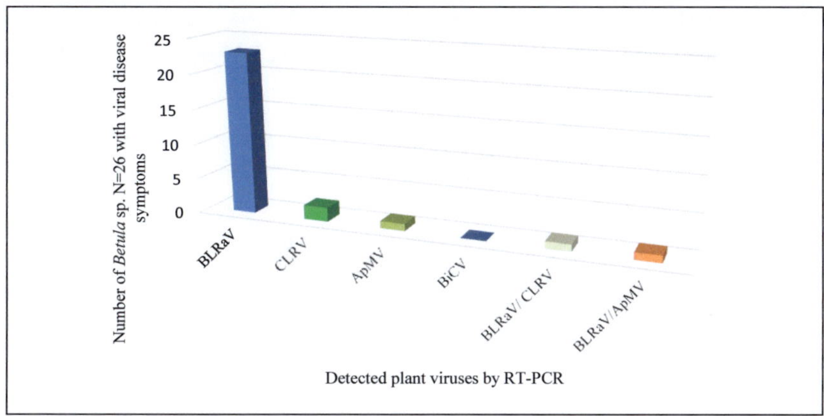

Fig. 26 Single and mixed viral infection detected by RT-PCR; single infection and mixed viral infection (two viral species infection) -Rovaniemi (Finland) June, 2017.

3.6 Serological results

Based on the detection of CLRV in bioassay leaf samples by RT-PCR, there was the need to confirm CLRV by applying an alternative serological approach. ELISA was chosen due to the availability of CLRV antibodies and the easy performance of ELISA. Both results were compared. Most of the samples tested with BIOREBA CLRV-Ch/DAS ELISA had CLRV infection whiles that of BIOREBA CLRV-e/DAS ELISA recorded little or no CLRV infections indicating that the cherry isolate was more prominent in the investigated trees. Nine symptomatic leaves which were collected from different part of a specific tree (TT SEAL 0407510) (E-54094) were infected with CLRV-ch (Fig. 27).

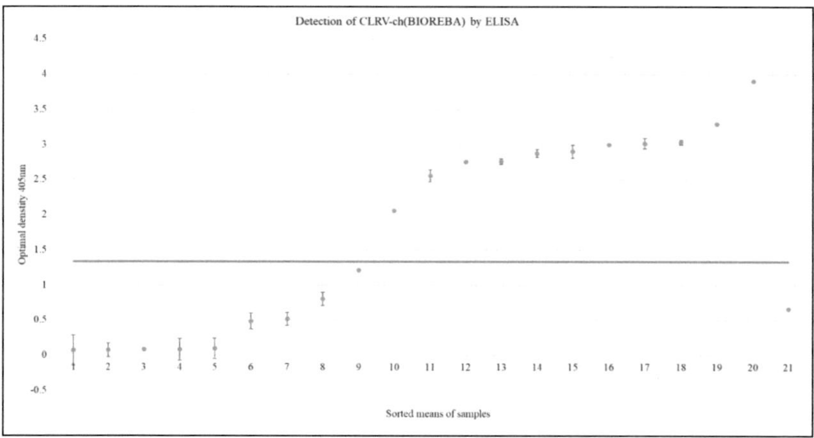

Fig. 27 Detection of CLRV by DAS ELISA in *Betula sp.* using antibody developed against cherry isolates of CLRV (BIOREBA, Art. No: 150915). Samples above the cut-off represent positive results and below the cut-off line represented negative tested results. Leaf sample numbers (1-20); (21= buffer) are listed in Appendix 11.3; Tab. A26. For dots without arrow bars, numbers were small. Calculations were done based on instructions from BIOREBA. (Sample 18 = Positive control from BIOREBA), the mean of three replications from each leaf material were included in calculations of thresh hold. The thresh hold was calculated using the formula «mean value + 3 x standard deviation + 10%».

Three birch leaf samples with numbers (407553BpenMO291B, 407554BpenMO291D, 407555BpenMO291F) known to be infected by CLRV (personal communication from Cutler, 2016) confirmed by RT-PCR had positive infection of CLRV-ch but not CLRV-e and E-54095. The results from all tested *Betula* leaf samples by CLRV-e antibodies had 2 samples testing positive (Fig. 28).

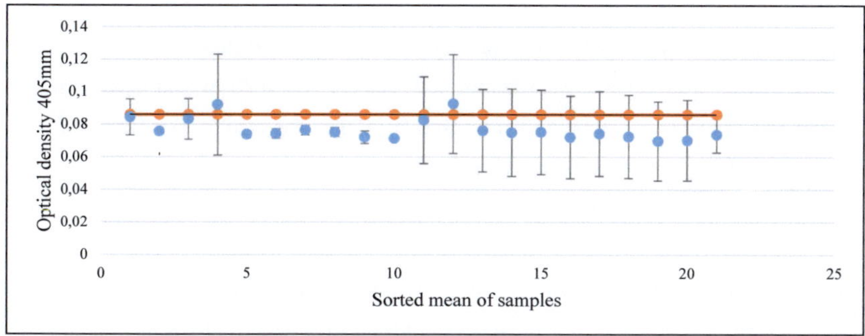

Fig. 28 Detection of CLRV by DAS ELISA in *Betula sp.* using antibodies developed against an elderberry isolates of CLRV (BIOREBA 150953). Samples above the cut-off represent positive results and below the cut-off line represented negative tested results. Leaf sample numbers (1-20); (21= buffer) are listed in Appendix 11.3; Tab. A26. For dots without arrow bars, numbers were small. Calculations were done based on instructions from BIOREBA. The mean of three replications from each leaf material were included in calculations of thresh hold. The thresh hold was calculated using the formula «mean value + 3 x standard deviation + 10%».

3.7 Database analysis of the *Betula* genome-NCBI

The bioinformatic analysis of the *Betula* sp. after BLASTx of BLRaV genome *(H0005 Bet-Germany-2014a-BLRaV 2,682 reads from H0005-Paired Reads mapped to BLRaV-Bet-Finland-2014b-AR (NC-040635)* against the database whole genome shotgun sequence (WGS) resulted in the extraction of partial sequences belonging to the group of badnavirus similar to BLRaV. Some results from BLASTn and BLASTx of BLRaV genome sequence using NCBI after filtering is shown in Appendix 11.4 (Fig. A8).

3.8 Biodiversity of plant viruses in *Betula* sp. -Whole genome shotgun (WGS)

Results from WGS database after BLASTn of the BLRaV sequence against the organisms; *Betula* (taxid: 3504), *Betula nana* (taxid: 3505), *Betula pendula* (taxid: 3505), *Betula verrucosa* (taxid: 3505) and *Betulaceae* (taxid: 3514) produced several contigs (Appendix 11.4; Fig. A6 and Fig A7).

3.8.1 Quest for integration of BLRaV-like sequence in the genome database of *Betula*

The sequence of BLRaV with an accession number MG686420 was used to find matching sequences in the whole genome shotgun database using BLASTn. The results after Blastn of the genome sequence (BLRaV) against whole genome shotgun are shown in (Fig. 30). Contig 1188 with an accession number FXXK01001726.1 was chosen after filtering based the percentage identity which was above 70% with a significant similarity to BLRaV. The different ORFs downstream and upstream of area of similarity to BLRaV were evaluated using BLASTn and BLASTx regarding their function. (Tab.17) shows results of the functional analysis of the downstream and upstream ORFs.

```
RID: X7XEF5ZN015
Job Title:H0203_Bet_Finland_2016a 1,737 reads from H0203_Paired...
Program: BLASTN
Query: H0203_Bet_Finland_2016a 1,737 reads from H0203_Paired Reads ( mapped to BLRaV_Bet_Finland_2014b_AR (NC_040635) using Geneious ID: lcl|Query_44427(dna)
Database: wgs (2 databases)

Sequences producing significant alignments:
                                                        Max     Total Query   E    Per.
Description                                             Score   Score cover Value Ident  Accession
Betula nana WGS project CAOK00000000 data, contig 62091, whole... 86.9   86.9  1%    9e-14 76.61  CAOK01062084.1
Betula nana WGS project CAOK00000000 data, contig 4687, whole...  86.9   86.9  1%    9e-14 76.61  CAOK01004687.1
Betula pendula genome assembly, contig: Contig519, whole genom... 63.5   172   0%    1e-06 87.04  FXXK01000520.1
Betula nana WGS project CAOK00000000 data, contig 340, whole...   63.5   63.5  0%    1e-06 84.21  CAOK01000340.1
Betula nana WGS project CAOK00000000 data, contig 17115, whole... 59.9   59.9  0%    1e-05 83.64  CAOK01017113.1
Betula nana WGS project CAOK00000000 data, contig 8411, whole...  59.9   59.9  0%    1e-05 78.57  CAOK01008411.1
Betula pendula genome assembly, contig: Contig1725, whole geno... 59.0   158   0%    1e-05 85.19  FXXK01001726.1
Betula pendula genome assembly, contig: Contig1329, whole geno... 59.0   59.0  1%    1e-05 71.64  FXXK01001330.1
Betula pendula genome assembly, contig: Contig1188, whole geno... 59.0   109   1%    1e-05 71.64  FXXK01001189.1
Betula pendula genome assembly, contig: Contig2691, whole geno... 56.3   110   1%    2e-04 70.63  FXXK01002692.1
```

Fig. 29 Blast result from BLRaVsequence *(H0005_Bet_Germany_2014a_BLRaV 2,682 reads from H0005_Paired Reads (mapped to BLRaV_Bet_Finland_2014b_AR (NC_040635)* using Geneious) with focus on *Betula pendula* genome contig1188 after blasting against whole genome shotgun. Maximum score =59.0, total score=109, E=value 1e-05, percentage identity 71.64% and accession number =FXXK01001189.1 (NCBI, 2021).

The blast from the BLRaV sequence *H0005_Bet_Germany_2014a_BLRaV 2,682 reads from H0005_Paired Reads (mapped to BLRaV_Bet_Finland_2014b_AR (NC_040635)* using Geneious resulted in some significant genome of *Betula* shown in (Fig. 30). After filtering the *Betula* genome with contig 1188 with a percentage identity of (71.64 %) was chosen because they gave significant results similar to BLRaV with possible function of the virus. Result obtained from BLASTn was displayed in the related *Betula* genome region (contig) using graphical overview of the sequence (Format-graphics) from NCBI (Fig. 30).

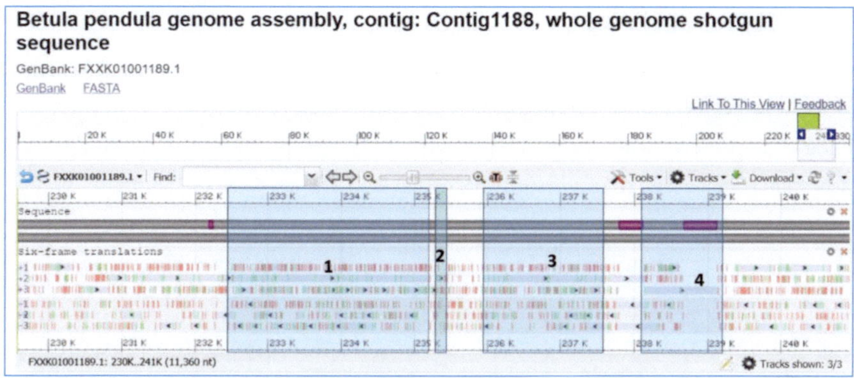

Fig. 30 Graphics of *Betula pendula* genome assembly with contig 1188 with forward and reverse translations. Regions highlighted in green are the start codons and red are the stop codons; regions highlighted in grey are the ORFs. For six frame translations, the forward translation reads from +1, +2, +3 and reverse reads from -1, -2, -3. The region of interest similar to BLRaV was the region (3) which seems to have protein function of Reverse transcriptase and putative RdRp as shown in blue colour (Tab. 18).

Tab. 18 Results after BLASTx and BLASTn of *Betula pendula* genome assembly with contig 1188 with forward and reverse translations. Regions highlighted with numbers 1, 2, 3 and 4 represents various range of OFRs which were Blasted via NCBI and resulted in hit from Badnaviruses. Target position (1) and position (3) gave significant results similar to BLRaV, Upstream and downstream was checked if there were Badnaviruses. The target position, E-value, accession numbers and results from BLASTn and BLASTx are listed in Tab. 18.

Target position	Accession number	Database nr focus taxid viruses (BLAST n)	E-value	Accession number	Database focus on viruses BLASTx	E-Value
(1) ORF (+2)	MT036049.1 NC_055598.1 KF545097.1	*Camellia-associated badnavirus isolate JX1, complete genome* *Camellia lemon glow virus isolate LG, complete genome* *Banana streak OL virus isolate Ke154 ORF III polyprotein gene, partial cds*	2e-08 2e-08 1e-06	YP_009666308.1 QPF16710.1 QBP37036.1	Reverse Transcriptase [Cladosporium fulvum T-1 virus] putative RNA-dependent RNA polymerase [Aedes aegypti virus 2] putative RdRP [*Lampyris noctiluca errantivirus* 1]	2e-162 9e-90 1e-87
(2) ORF (+2)		No significant similarity found.			No significant similarity found.	
(3) ORF (+2)	MG686420.1 NC_040635.1 NC_038995.1 KF545136.1	*Birch leaf roll-associated virus isolate BpubFin407501, complete genome (Badnavirus)* *Birch leaf roll-associated virus isolate BpubFin407507, complete genome (Badnavirus)* *Dioscorea bacilliform TR virus strain DBV9, complete genome (Badnavirus)* *Banana streak UA virus isolate Ug228 ORF III polyprotein gene, partial cds*	0.001 0.001 0.001 0.001	YP_009666308.1 QPF16707.1 QBP37036.1	Reverse Transcriptase [Cladosporium fulvum T-1 virus] putative RNA-dependent RNA polymerase [Aedes aegypti virus 1] putative RdRP [Lampyris noctiluca errantivirus 1]	6e-124 3e-86 2e-77
(4) ORF (+2)		No significant similarity found.			No significant similarity found.	

3.8.2 Capillovirus-like sequence in the genome database of *Betula*

A search through a *Betula* genome database resulted in a contig 934 with accession number FXXK01000935.1. Further investigation provided significant hit related to Capillovirus-like sequence which seems to be integrated in the *Betula pendula* genome after upstream and downstream BLASTx of OFRs. Accession number MK402233.1 and MK402234.1 with E-value of zero, were detected in graphics of *Betula pendula* genome assembly with contig 934 (Fig. 31).

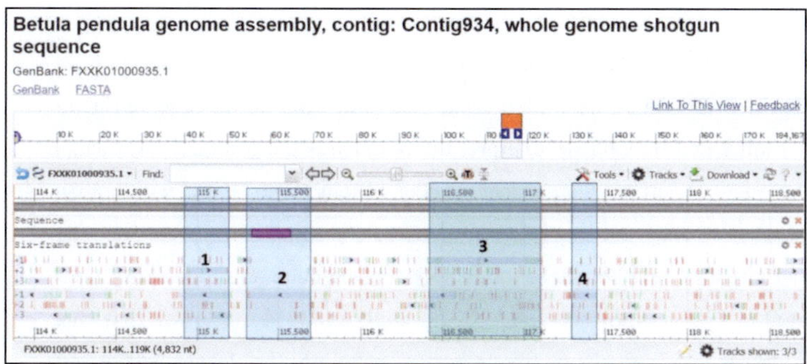

Fig. 31 Graphics of *Betula pendula* genome assembly with contig 934 with forward and reverse translations. Regions highlighted in green are the start codons and red are the stop codons; regions highlighted in grey are the ORFs. For six frame translations, the forward translation reads from +1, +2, +3 and reverse reads from -1, -2, -3. The region of interest similar to Birch Capillovirus was the region (1) and region (3) which seems to have protein function of polyprotein and coat protein as shown in Tab.19.

Tab. 19 Results after BLASTx and BLASTn of *Betula pendula* genome assembly with contig 934 with forward and reverse translations. Regions highlighted with numbers 1, 2, 3 and 4 represent various range of OFRs which were blasted using NCBI and resulted in hit from Birch Capillovirus. Region (3) gave significant results similar to Birch Capillovirus, Upstream and downstream was checked if there were Birch Capillovirus. The target position, E-value, accession numbers and results from BLASTn and BLASTx are listed in Tab. 19.

Target position	Accession number	Database nr focus taxid viruses (BLAST n)	E-value	Accession number	Database focus on viruses BLASTx	E-Value
(1) ORF (+2)	-	function unknown No significant similarity found.	-	QHB15185.1	polyprotein [*Endogenous retrovirus sp.*]	6e-05
				AAD37341.1	Reverse transcriptase [*Cauliflower mosaic virus*] polyprotein 1 [*Petunia vein clearing virus*]	1e-04 2e-04
(2) ORF(-1)	-	function unknown	-	-	function unknown	-
(3) ORF (+1)	MK402233.1	*Birch capillovirus isolate BpenGer407526-B5 polyprotein gene, partial cds*	0.0	QDX18417.1	Polyprotein [Birch capillovirus] polyprotein [Birch capillovirus]	1e-98 4e-69
	MK402234.1	*Birch capillovirus isolate BpubFinn407501-3A polyprotein gene, partial cds*	8e-131	QDX18418.1 ACI46655.1	coat protein [*Citrus tatter leaf virus*]	2e-23
(4)ORF (-1)		function unknown			function unknown	

4. Discussion

Plant viruses are widespread in all plants and cultivars worldwide and in many cases do have a strong impact. But which viruses do we find in urban trees and what do we observe due to their impact on tree vitality and how do we interpret the findings is the central focus of this thesis with particular view to urban *Betula* species. For a long time, there was human perception that, viral diseases may not be present in deciduous trees, but research since the late 80th suggests that viruses could be one of the factors causing decline of deciduous trees (Nienhaus and Castello, 1989; Büttner *et al.*, 2022). Hull, in 2014 observed that the interaction between viruses and host plants affects host morphology (i.e. leaves, stems and fruits etc.) and physiology negatively, resulting in disease. Plants do interact with many threats such as viruses which change the metabolism of plants without obvious symptoms in the beginning over a longer time.

Unlike other pathogens such as fungi and bacteria which are visually identified by presence of significant characteristics (e.g. fungal hyphae or spores), viruses are however not visible to the naked eye (Agrios, 2005). They can instead be observed with the aid of transmission electron microscope. From forestry, it is known that climate change, pest and pathogen damage, land-use changes as well as forest fragmentation may affect trees and cause reduction in genetic diversity making the forest more vulnerable to other threats (Gauthier *et al.*, 2015). Abiotic factors such as road salt, natural gas, nutrient toxicity and deficiency, soil salinity and acidity, and radiation causes trees to become weak and susceptible to diseases and pests (Teshome *et al.*, 2020). These factors to some extent could also be applied in green urban trees.

Betula sp. (*B. pendula* Roth. and *B. pubescens* Ehrh.) known to be a pioneer tree have major impact to the ecosystem and an economic importance in the industrial production of paper and plywood (Dubois *et al.*, 2020), but during the last two decades we have become aware that virus infection can seriously affect trees. Due to its pioneer tree properties, birch can withstand the extreme conditions in the urban environment and therefore was long-time favoured for planting in towns and cities. The recommendation of birch for planting in cities changed in the last decade (GALK Straßenbaumliste, 2022). As birches might suffer from Birch leaf roll disease (BLRD) as we know e.g. from the metropolis Berlin (Rumbou *et al.*, 2018). Trees have difficulties to withstand other factors like pathogens and changing climate towards heat and drought, hence the planting of birches in streets of Berlin has been stopped.

With a total of about 438,159 deciduous trees in Berlin, *Betula* sp. represents 3% in urban green space (Senatsverwaltung Berlin, 2016). About 997 diseased trees were felled in the year 2015, with the *Betula* sp. being the third most felled tree species with around 80 trees (8 %; Anonymous: BA Fachbereich Grünflächen, 2016). There seems to be a drastic decline of trees from 2016 to 2019 (BUND Baum report, 2021). Although the decline has been attributed to abiotic factors (i.e. temperature and drought), a wide reliable survey on biotic factors has not been ruled out so far. *Betula* sp. grown in streets of Berlin cannot tolerate drought and excessive water in their roots and eventually die because of their inability to build up and store any reserves substances (Tidow, 2020). Due to climate change (i.e. high stress impact by drought and temperature) especially in summer, the survival rate of birches is limited due to low precipitation. The annual low precipitation (with a long-term average of 598 mm) in Berlin and uneven distribution of precipitation over the year (2018 and 2019) impairs the vitality of trees, exposing them to harmful organisms such at the insect *S. ratzeburgii* (Tidow, 2020).

The report from the Berlin Plant Protection Office shows that, persistent drought had effect on street trees especially in spring and summer months since 2018 (Pflanzenschutzamt Berlin; Deutscher Wetterdienst, 2020). Unavailability of water in spring affects birches, because they need water to inflate and sprout their leaves that were in the buds the previous year (BUND Baum report, 2021). A report by Possen *et al.*, 2011 where three silver birch genotypes were exposed to three soil moisture conditions (wet-control-dry) showed that, drought stress reduced the growth, biomass, shoot-to-root ratio and gas exchange, while the wet treatment had opposite effects. Due to small soil planting holes which are surrounded by asphalt along streets of Berlin, it is difficult for rains to reach the roots of birches. This water needed by the roots instead run straight through sealed areas into sewage system leading to drought and tree damage (BUND Baum report, 2021). In addition to that, temperatures in the city of Berlin have been reported to be warm due to lack of air exchange, heated streets and house facades (Tidow, 2020). This could easily influence the birches since they are sensitive to higher winter temperatures in some species and may lead to delayed bud break which may need more time to have low temperature requirement fulfilled (Junttila *et al.*, 2012).

Biotic stresses such as insects, fungi, nematodes and viruses take advantage of damaged trees once they become weak leading to considerable losses. For example, the fungal pathogen *Melampsoridium botulinum* has been reported to affect *Betula* sp. causing stunted growth and reduction of life expectancy (Beck *et al.*, 2016).

The appearances of these bark beetles which causes severe damage to *Betula* sp. over the last years (2018 and 2019) are stimulated by warm and dry conditions in Berlin (Tidow, 2020). Under favourable growth conditions, e.g. fungi can also cause extensive wood decay in relatively short time which destroys the cell wall components (i.e. cellulose and lignin) as well as conductive tissues (Downer, 2019). Stress from wind, heavy rain as well as other abiotic factors may cause trunks and branches of trees to fall off due to insect and fungi interactions which are normally visible on birches.

Although *Betula* sp. is resistance to pest and diseases, climate change and globalization could increase insect damage to birches (Heimonen *et al.*, 2016). With a mean average temperature needed for survival of birches, increase in temperature could lead to generation of herbivorous insect which can have adverse effect on birch growth and productivity. During warm winters, birches are able to survive by going to state of dormancy and forms bud immediately in early spring once conditions are favourable. In general, disease leading to decline of tree population involves a complex interaction between multiple biotic and abiotic stresses (Teshome *et al.*, 2020).

A recent article reported that, the degree of virus accumulation in its host is likely to be expressed seasonally because both viral replication and plant growth are dependent on temperature (Honjo *et al.*, 2020). Plant viruses such as *Cherry leaf roll virus* (CLRV), *Birch leaf roll associated virus* (BLRaV), *Birch carlavirus* (BiCV) and *Apple moasiac virus* (ApMV) were detected in birch leaf samples collected in summer (May, 2017) with an average temperature of 20°C and corresponds to reports that; viruses can replicate at an optimal temperature range from 15 °C and 30 °C (Jones *et al.*, 1990; Ohsato *et al.*, 2003; Chung *et al.*, 2016).   It is however not clear if viruses infecting birches can be detected in bark of trees and roots due to unavailability of leaves in winter. Apart from plant viruses infecting leaves and other tissues, the *Beet necrotic yellow vein virus* (genus Benyvirus) has been reported to be one of the soil-borne viruses causing disease symptoms in roots of sugar beet (Biancardi and Tamada, 2016) but roots from virus infected birches have so far not been studied. It is unknown if these viruses detected replicates or are dormant in winter and becomes active during spring and summer.

Viral infection on both herbaceous and deciduous trees follows natural dynamics and has peaks of abundance. It can be argued that replications of viruses reach their peak, symptoms are extra ordinarily high during summer hence the detection of viruses in streets of Berlin.

The birch leaf symptoms are seen more often in the spring (Personal communication with Landgraf) when temperatures are relatively low and the plant becomes active. Intercostal chlorosis being the most prevalent in Berlin, Rovaniemi and Wildberg/Nagold gives a hint of chloroplast-virus-interaction even though it was not investigated. The chloroplast has been reported as a common target of plant viruses (Zhao *et al.*, 2016) and viruses could have influence on the chloroplast from birch leaves via replication. For example, typical symptoms of chlorosis have been reported to be attributed to chloroplast virus-interaction and dependent on factors such as; changes in chlorophyll fluorescence and reduced chlorophyll pigmentation (Balachandran *et al.*, 1994b), inhibited photosystem efficiency (Lehto *et al.*, 2003), imbalanced accumulation of photo assimilates (Lucas *et al.*, 1993; Olesinski *et al.*, 1995, 1996; Almon *et al.*, 1997). Changes in chloroplast structures and functions (Bhat *et al.*, 2013; Otulak *et al.*, 2015), and repressed expression of nuclear-encoded chloroplast and photosynthesis-related genes (CPRGs) (Dardick, 2007; Mochizuki *et al.*, 2014a) and direct binding of viral components with chloroplast factors (Bhat *et al.*, 2013; Zhao *et al.*, 2013) has also been reported.

However, some birch leaves in Berlin exhibited no symptoms but viral infection was detected within. It could be that not all viruses could influence chloroplast, cell division and leading to deformation. On the other hand, chloroplast factors (i.e. salicylic and jasmonic acid) seems to play active roles in plant defense against viruses (Heiber *et al.*, 2014) even though it is poorly studied. Viral infections may cause changes in the morphology of chloroplast and cell metabolism, which may have adverse effect on plant growth (Hamacher, 1994). The observation regarding viral symptomatology made in this study can support this way of interpretation. Especially the vein banding was correlated to a later appearing necrosis and final loss of tree crown. Infected trees tended also to hold dead branches and crown dieback. But what is known about the biological side effect of symptom pattern like obvious yellow ring spots, variegation or line pattern? Insect vectors such as aphids causes damage to plant directly by feeding and indirectly by transmitting viruses to wounded places, from where viruses spread throughout the plant (Wielkopolan *et al.*, 2021).

Some insect vectors have been reported to be sensitive and select plant host based on sensory cues such as leaf colour, olfactory (emission of volatile organic compounds) gustatory or tactile stimuli (Heard, 1999; Mauck *et al.*, 2014). Plant pathogens, including viruses, induce changes in plant phenotypes, their palatability, and nutrients components, to enhance attraction of plants by insect vectors and to increase pathogens acquisition and transmission to other plants (Lieutier *et al.*, 2009; Chesnais *et al.*, 2020).

This confirms to what was observed in samples from one of the stands in Berlin (Rudow) due to presence of insect population in such region. Symptomatic birch leaves were infected by ApMV hence; we could conclude that ApMV could have probably induced phenotypical changes causing attraction of insects as described by Gutiérrez *et al.*, 2013. Having detected ApMV in birches, it is unknown if ApMV which causes huge losses in fruit trees like *Prunus* species and Apple will have same effect on birch trees due to it wide host range as reported by Kinoti *et al.*, 2018.

Over many years, virus infected birch trees in streets and parks in Berlin have been studied. There has been a gradual decline of *Betula* sp. which seems to be associated with viral diseases such as *Birch leaf roll associated virus* (Rumbou *et al.*, 2018), *Cherry leaf roll virus* (Büttner *et al.*, 2011; von Bargen *et al.*, 2009; Jalkanen *et al.*, 2007; Jones *et al.*, 1990; Cooper and Atkinson, 1975), *Apple mosaic virus* (Bandte *et al.*, 2009; Cooper and Massalski, 1984), *Arabis mosaic virus* (Bandte *et al.*, 2009; Polák and Zieglerová, 1997; Polák and Procházková, 1996; Hardcastle and Gotlieb, 1980; Gotlieb and Berbee, 1973), *Tobacco necrosis virus* and *Tomato ringspot virus* (Cooper and Massalski, 1984).

The view on *Betula* sp. over the years has given a remarkable variety of leaf symptoms such as intercostal chlorosis, ringspot, interveinal chlorosis, line pattern, leaf blistering, oak leaf pattern, leaf mosaic and leaf variegation as reported by Opoku *et al.*, 2018. The variety of leaf symptoms with most of them being different forms of chlorosis were categorized into various group as described by Cooper and Massalski 1984. Studies over the last years on leaf symptoms including the whole crown of virus infected birch trees from several stands in Berlin indicated a variety of symptoms. Due to complex diversity of *Betula* leaf symptoms, it is very difficult to correlate observed symptoms to the causative plant viruses. Different symptoms keep recurring each year, but sometimes it is difficult to observe the same symptoms each year due to interactions between host, virus and other mimicking biotic factors (e.g. insect bites and fungi spores). As these reoccurring symptoms in birches are very similar to virus associated symptoms in other plants, the conclusion of viral involvement in the formation of these symptoms was drawn by Division of Phytomedicine -HUB (Bandte *et al*, 2022, Rumbou *et al.*, 2018).

Nonetheless, biotic stresses caused by other living organisms (pathogens, nematodes, insects, arthropods) as well as abiotic stresses (temperature and moisture extremes, mechanical damage, chemicals, nutrient deficiencies or excesses, salt damage and other environmental factors) are all interacting factors which have to be evaluated individually for each stand and tree species.

68

Although some tree species can withstand drought stress more readily than others, severe water deficit due to increased temperatures can lead to typical dieback symptoms and branch decline of trees such as the *Fagus sylvatica* (Walthert *et al.*, 2021). Vasquez *et al.*, 2022 reported that temperature and drought generate stress in potato plants which induce symptom expression by suppressing the salicylic pathway, and increases *Potato yellow vein virus* (PYVV) replication. Although, there is little investigation on virus-drought stress interaction on woody trees, viruses detected in birches might also replicates due to drought and temperature.

The infection of *B. pendula* by CLRV was first reported in Germany and Great Britain (Schmelzer 1972; Cooper and Atkinson, 1975) and investigated by mechanical inoculation of plant saps extracted from leaves and roots of *Betula* sp. onto herbaceous host and serological detection by ELISA. Although variety of symptoms were observed in this study, chlorotic ringspot and leaf roll are reported to be correlated with CLRV based on assays carried out successfully by (Nienhaus *et al.*, 1985). During that time, other virus (e.g. BLRaV, ApMV, BiCV, *Birch caulimovirus* and *Birch Capillovirus* were not known and detection methods were limited to serology and electron microscopy. Mechanical transmission was the means of detection and this could not check for contamination by additional viruses and whether antibodies produced against CLRV had also targeted mixed infections. Mixed infection by several viruses was difficult to investigate due to the limitations in diagnostic tools during that time. Consequently, symptoms were addressed mainly to CLRV. But variety in symptoms give a different picture and one could have asked how all those symptoms can be created by a single type of virus.

With the results compiled in this thesis, there were *Betula* leaves exhibiting symptoms of chlorotic ringspot, vein banding and leaf roll which were associated with *Birch leaf roll associated virus* (BLRaV). The results from this thesis confirmed the presence of CLRV in birch tree species exhibiting viral symptoms from Rovaniemi (Finland) using RT-PCR but not (immune capture-RT-PCR) as reported by Jalkanen *et al.*, 2007.

The diversity of the foliar symptoms was confirmed in this study but was finally addressed to a complex situation of coinfection by several viruses which in contrast with what was reported by (Bandte and Büttner 2001; Büttner *et al.*, 2013) where CLRV was associated to foliar symptoms from chlorotic vein banding, chlorotic ring spots, mottling, necrosis, leaf roll and reduced leaf size to dieback of branches or whole tree.

Also birch specific variants of CLRV (with accession number MK402281 and MK402282), has been published recently within this viral complex (Rumbou *et al.*, 2020). CLRV strains vary and are specific for host species (e.g. CLRV from walnut, hazel nut, cherry and rhubarb). CLRV infecting birches from streets of Berlin, 2017 were randomly distributed and this is in confirmation with what was reported by (Rebenstorf *et al.,* 2005; Büttner *et al.*, 2011) with wide natural host range with 28 species of broad-leaved trees, shrubs and herbaceous plants causing severe effects on them. CLRV and its variants are mainly reported from *Betula spp, Sambucus nigra, Juglans regia* and *Prunus avium, Betula* spp (Büttner *et al.*, 2013). Even though some insects were observed during field sampling especially in Unter den Birken and Kopenick in Berlin 2017, there was much focus on birch leaf with symptoms. These insect could probably be one of means of CLRV transmission as reported by Büttner and Koenig, 2013. Their mode of transmission is documented via seeds, pollen, vectors (e.g. insects), mechanically by roots intergrowth and water (Büttner and Koenig, 2013). CLRV can be transmitted via insects even in an urban environment, but little is known regarding potential vectors in urban environment.

A case study on potential virus vectors in birch trees in Berlin confirmed the *Kleidocerys resedae*, heteroptera to be the largest and most prevalent group of insects (43%) on *Betula pendula* (Langer *et al.*, 2013). Ten bug species, eight *cicada* species, and nine aphid species were determined in CLRV-infected birches (Langer *et al.*, 2013). *Kleidocerys resedae* and *Polydrusus sp.* tested positive for CLRV and could be a potential vector for transmission of CLRV. Apart from vector transmission, CLRV is reported to be transmitted by seed and pollen (Rebenstorf *et al.*, 2006). The option of virus transmission by pollen feeding insects has to be taken into consideration and if possible investigated. A variety of CLRV strains reported on Corsican *Betula* sp. from Montane and Mediterranean origin by Langer *et al.*, 2016 provided a first insight into CLRV birch population that has maintained within an isolated population. With the same primer combination (CP350F/CP977R and CP188F/CP350R) provided by Langer *et al.*, 2016, CLRV was confirmed in birches from Berlin.

Investigations on the spectrum of genetic variants occurring in individual CLRV infected trees and a comparative estimation of this parameter gives an idea on the value of genetic adaption during coevolution of CLRV in different host species in a specific environment as well as a chronological course of CLRV infection events in a specific geographic region (Langer *et al.*, 2016).

The HTS results from the birch database confirmed the presence of new novel viruses such as Birch capillovirus, Birch caulimovirus which confirms reports from von bargen *et al.*, 2009 that other viruses could be involved in the disease affecting birches. Before HTS, the emerging phenomenon described as Birch leaf roll disease (BLRD) was associated to the presence CLRV in the affected trees based on standard molecular diagnostic tools (von Bargen *et al.*, 2009; Rumbou *et al.*, 2016). However, HTS revealed the presence of BLRaV in affected birches (Rumbou *et al.*, 2018) which indicates that the BLRD could be influence not only CLRV but BLRaV as well as other unknown viruses. New strains of BLRaV infecting birches from Corsica and Berlin were confirmed in this study as reported by Pack *et al.*, 2019. The variety of viral contigs established from HTS gave knowledge of presence of much more viral species infecting birches (Landgraf *et al.*, 2018).

The HTS analysis workflow was reorganized (Köpke, 2019), and structured. Software like Geneious prime was implemented including tools like Spades for de novo assembly of HTS reads and updated which resulted in more virus related contigs in the datasets than could be identified previously by commercial providers like baseclear. Intensive studies on selected *Betula* stand in Berlin, Wildberg/Nagold and Rovaniemi (Finland) gave hints of more viruses such as BLRaV, birch caulimovirus, birch capillovirus and birch idaeovirus involved than previously assumed. For example, BLRaV which is introduced in details in chapter 4.2.1.

4.1 Selected *Betula* stands -Berlin -Germany

The monitoring of *Betula* stands in Berlin (May, 2017) resulted in the observation 16 different symptoms which is in contrast with what was observed in the year 2015/2016, where 14 different symptoms were reported (Landgraf *et al.*, 2016). Although there were similarities in symptoms observed in both years, the diversity of symptoms in different years in summer (2015/2016 and 2017) represents an individual result of the person doing the visual rating since it is very difficult to see the whole tree crown of a particular tree. Different symptoms such interveinal chlorosis, necrosis and ringspot, observed during the study could be attributed to time of year of investigation (season), the nutrient conditions of soil, water availability and weather or climate conditions in the different stands.

According to Ahamedemujtaba *et al.*, 2021, symptoms caused by *Piper yellow mottle virus* (PYMoV) in black pepper were severe with increase in temperature and relative humidity. This could be the reason for symptoms of intercostal chlorosis in birches since samples were collected in (May/June 2017) with increased temperatures and drought as reported by (Tidow, 2020) in Berlin.

On the other hand, virus symptoms may disappear temporarily under certain environmental conditions (e.g. high or low temperature) and such symptoms are described as masked. Birch leaves with masked symptoms observed in this study had viral infections. Constable *et al.*, 2013 reported that, viral symptoms of grapevine appeared in older leaves with severe symptoms of chlorosis and but disappeared in summer. Hence, this could be the reason for detection of viruses in some of the masked birch leaves investigated. In contrast to that, Geminiviruses for example produces severe symptoms in Cassava at low temperatures and less severe in rising temperatures (Chellappan *et al.*, 2005).

A very distinct symptom with double ringspot, double oak leaf and mottling was first observed in Rudow (A; E56736) and Reinickendorf (E; 54085) respectively. Although both leaf samples (A; E56736 and E; 54085) were associated with *Apple mosaic virus* (ApMV), other viruses cannot be excluded. Ringspot symptoms is reported to be associated with ilarviruses group (ApMV) (Shiel *et al.*, 1995), hence it was obvious these symptoms with double ringspots from birches had ApMV infections. By comparison of databases, certain similarities of strains from Rudow to ApMV strain from Hazelnut tree were detected. It is still unknown, why infection by different strains of the same virus (e.g. ApMV) results in the development of a variety of symptoms or even an enhancement. Synergistic or antagonistic interplay of ApMV and other plant viruses such as BLRaV could be the reason for symptom diversity since ApMV and BLRaV recorded significant amount for mixed infection.

Virus infections are often confused with symptoms caused by other biotic and abiotic factors, and these have to be distinguished by reliable diagnostic methods (Bandte *et al.*, 2022; Büttner *et al.*, 2013). For samples in Berlin, different forms of chlorosis such as ringspot, oak leaf pattern, leaf roll, vein banding, mottling with ringspot, chlorotic spot, vein chlorosis, and vein banding, were observed and could be attributed to multiple factors. Necrosis observed during sampling was obviously a symptom from drought or nutrient deficiency but plant viruses were however detected in them.

Some leaves also had spots from insect bites forming moulds and holes in leaves. Symptoms on birch leaves are known to spread over whole leaf blade having begun from one focal point (Büttner *et al.*, 2013).

72

In this study, young leaves were often found to be deformed or involute, which supported these preliminary findings. The symptomatology in birch leaf reported by Büttner *et al.*, 2013 such as mosaic like leaf patterns, ringspots, lines and mottling was reconfirmed in this study. However, symptoms of double ringspot, leaf roll, vein banding and necrosis were additionally observed during the survey. Viral symptoms have to be distinguished from developmental stages and the phenology of special phenotypes. Although birch symptoms were associated to viruses, they might also be attributed to alterations hormones or metabolism at the endogenous levels of plant growth substances as reported by Culver and Padmanabhan 2007; Lewsey *et al.*, 2009.

Symptomatology is often influenced by personal interpretation; therefore, future concern should also focus on computer based tracking system for categorization of symptoms. With the development of computer vision digital learning, progress has been achieved on recognition and diagnosis of some plant diseases. Mahlein *et al.*, 2012, reported that Hyperspectral imaging (HSI) offers high potential as a non-invasive diagnostic tool for sugar beet leaf diseased with Cercospora leaf spot, powdery mildew and leaf rust at different developmental stages. A similar device could also be developed for easy identification of symptoms and related viruses in trees like birches. Viral symptoms have to be distinguished from developmental stages and the phenology of special phenotypes.

Soil nutrition and texture, weather conditions, ultraviolent index, water accessibility and many other environmental and cultural conditions influence the overall health of a tree as well as the symptomatology (Boa, 2003). Insect damage such as (chlorotic spots) can sometimes be confused with plant diseases caused by microorganisms or abiotic factors. Insect damage can open wounds for viral transfer and a new infection. The overlapping symptoms, insect bites, moulds, could end up creating wounds in cell thereby enabling easy access of viruses to interact and replicate (Agrios, 2005). Chlorotic spots from insect damage were found in birch trees infected by BLRaV, but it is still too early to draw conclusions from this situation towards a natural vector transmission of BLRaV. Since BLRaV represents a new member of the genus Badnavirus, further research especially on mealy bugs with group of vectors of tropical Badnaviruses is necessary to confirm vector transmission for BLRaV. In temperate regions, different types of scale species *Lepidosaphes ulmi* (Oystershell scales) and *Quadraspidiotus perniciosus* (San José scale) are known to exist in birches (Kaur, 2022). As they belong to the group of *Coccinea* just like the tropical mealy bugs, there are at least some candidates for natural transmission of BLRaV in birch. From the tropical mealy bugs, it is known, that they end up taking badnaviruses via sucking of phloem (Bhat *et al.*, 2020).

Whether the BLRaV is mainly distributed in the phloem remains to be clarified, because it would have consequences for the mode of transmission and search of potential vectors. Knowing a complete history of the birch tree in combination with consequent observation and rating is essential to end up with an accurate diagnosis.

A plant specimen in the early stages of disease development should be examined to have a chance for treatment. Research is demanding the optimization of growth conditions for trees living under climate stress in urban environment (Salmond *et al.*, 2016). This includes the addition of water and nutrients as well as treatment by natural plant stabilizers. Once a tree has decayed, secondary organisms invade the tissue and evidence of the primary pathogen is often obscured. Although the leaf samples from Berlin are discussed in this chapter, the results from Wildberg/Nagold where tree nurseries were produced in clones had infections of BLRaV which could be a starting point of plant viral infection as reported by Roberts *et al.*, 2020.

Street trees from Berlin with BLRaV infections was very prominent in all areas where samples were collected but it is however unknown if there were already infected on the tree nursery site before being transferred to streets for planting. Again, Robert *et al.*, 2020 reported that, tree pathogens can spread to forest unit and urban streets via streams, wind, fog, animal (e.g. insects, feathers, and fur) or human vectors because they do not have spatial proximity connectivity. Based on this, the primary infection of birches could have begun from the forest via dispersal mechanism or human vectors.

Felled trees (stumps) as a result of thinning or partial cuttings were observed in some stands (e.g. Grünau and Köpenick) and this may be a source of primary infection of viruses. Although these stumps are important for forest biodiversity it may act as a reservoir for many pathogens and pests which could expose newly planted trees to inoculum (Hartley, 2002). Stumps can also provide a nutrient source for inoculum of wood decaying fungi with saprotrophic or even plant virus vector activity. Some knowledge on the transmission of viruses by fungi is given from agriculture e.g. from rhizomania disease of sugar beet (McGrann *et al.*, 2009) but little is known on the transmission of viruses by fungi in woody plant even though there are so many wood decaying and plant pathogenic fungi in trees (Rumbou *et al.*, 2020). Viral signatures from the group of Benyvirus are recognized in *Betula* HTS datasets from this study. Within the group of Benyvirus, transmission of viruses by fungi is known for some representatives. In response to this risk, site preparation may include stump removal or chemical treatment which can also have adverse effects on habitat management for the benefit of biodiversity.

Predisposing factors have to be taken into consideration. Viruses act as both predisposing factors weakening the plant as well as pathogen destroying the plant or attracting other pests depending on the aggressiveness of the viral species or strain as reported by Büttner *et al.*, 2013; 2022. The *Rice dwarf virus* (RDV) for instance can manipulate host to release volatiles that are either attractive or repellent to their insect vectors (Chang *et al.*, 2021). Since insect vectors were observed during survey, detected viruses could also influence olfactory behaviour of insect vectors causing them to move from one tree to the other. The vector Green rice leafhoppers (GRLHs) responsible for transmission of *Rice dwarf virus* revealed that, RDV infection significantly induced the emission of (E)-β-caryophyllene and 2-heptanol by rice plants influencing the olfactory behaviour of both non-viruliferous and viruliferous GRLHs. The virus may emit anti-feeding chemicals when a vector begins feeding on a plant to persuade the vector to switch to another plant or an uninfected plant (Chang *et al.*, 2021). On the other hand, beneficial viruses however do exist even though they have poorly been studied and unexploited in agriculture crops and forestry (Roossink, 2015).

4.2 Single and mixed infection of plant viruses detected by RT-PCR

80% of investigated symptomatic birch leaf samples in this study had either single or mixed infection from plant viruses (i.e. BLRaV, CLRV, BiCV and ApMV). This conforms to a complex viral population in grapevine (*Vitis* and *Muscadinia* sp.) reported to exist with mixed infections with different virus species or divergent variants of the same species (Martelli, 2017). The viral population in birches from Berlin seems to be even more complex based on the findings from this thesis. The molecular biological detection of plant viruses in diseased and degenerating birches from eight districts in Berlin have shown that single infection by BLRaV is widely distributed (27.8 %). Even though there were single infections from CLRV, ApMV and BiCV respectively, BLRaV was the most prevalent in all *Betula* samples investigated in 2017.

In contrast, previous study by Landgraf *et al.*, 2016 showed CLRV (71%) being more significant than the other viruses in the area of Steglitz -Zehlendorf (Berlin). The occurrence of these two different viral species (i.e. BLRaV and CLRV) in different years could be as a result of seasonal fluctuations, location of tree species as well as different geographical areas or different source of young trees. For instance, Beaver and Harper, 2021 reported that seasonal changes have adverse effect on *Tobacco ringspot virus* (TRSV) accumulation infecting *Vitis vinifera, Malus domestica,* and *Prunus armeniaca.*

Out of the four tested plant viruses, BLRaV recorded with 55.6% (40) as the most prominent virus in the investigated trees followed by CLRV with 26.4% (19), ApMV with 22.2% (16) and BiCV 8.3% (6) in leaves. The prevalent of BLRaV infection could be due to season or time of sample collection. Birch leaf samples were collected in summer (May-June) where average temperatures were recorded at 20°C. Rumbou *et al.*, 2018 confirmed presence of BLRaV in birch leaves around the same time in summer. Again Badnaviruses have been reported to be abundant in tropical and subtropical environment where warmth and humidity favours their major vectors such as mealy bug (Geering, 2021). This could be an obvious reason for detection of BLRaV in birches in summer (May-June, 2017) although the focus was not on mealy bugs.

In contrast to the study in 2015 and 2016, ApMV was detected in single infections as well as in mixed infections with a percentage (22.2 %) by RT-PCR. The difference in the detection of ApMV in 2015/2016 and 2017 might be due to regional or topographical differences in investigated areas, virus distribution in those areas and their specific plant origin or local climatic conditions. An extreme symptom with double ringspot and double oak leaf was first observed in Rudow, 2016 which conforms to the plum trees symptoms of ringspots and oakleaf with chlorotic line pattern associated with ApMV (Grimová *et al.*, 2013).

An additional proof of ApMV in the newly discovered double ringspot symptoms (Fig. 9; A; E56736) has been achieved by applying HTS (Contig number H0310 Bet_Rudow 2017). Severe symptom from this particular tree was however not observed in 2017 and 2021 respectively (Landgraf personal communication, 2021) since the branches had already died which might be due to fact that the viral life cycle showed up with this dynamic appearance of symptoms. Wells *et al.*, 1986 reported that, symptoms of *Prunus necrotic ringspot virus* (PNRSV-*Ilarvirus*) appear in the first year after infection of peach orchard and then become symptomless although some strains cause recurrent symptoms annually. This might be a possible hint for not observing symptoms in three consecutive years. The disappearance of the symptoms in 2017 and 2021 leaves a huge gap on whether lifecycle of the ApMV is dependence on the dynamic appearance of symptom.

Although ApMV (genus; *Ilarvirus*) have wide host range and differ in expression in different virus strains, some have relatively narrow range of natural host consisting of woody plants which may require several years after initial infection for the virus to infect trees uniformly. Having found ApMV in birches, it is possible that this virus can also have adverse effect on the tree by causing reduction of birch growth since the (Rudow strain; E56736) was highly similar to ApMV strain from hazel nuts.

Akbas and Degirmenci, (2009) reported that, nut cluster weight was reduced (on average by 42%) in ApMV infected hazelnuts, but nut size was also decreased with empty nuts. This might also affect the quality of seeds even though that hasn't been studied. However, trees seem to combat viral infection to develop a kind of equilibrium with its virome.

From my own observation, some older birch individuals (about 80 years) in urban green of Berlin were found to be infected by several viruses (virome) but still survive and produce flowers and fruits. For mixed viral infection in both years (2015/2016 -13.6% and 2017-12.5 %), there was no significant difference. However, the combination of CLRV and BLRaV seems to be prevalent in *Betula* trees and associated with the Birch leaf roll disease (Rumbou *et al.*, 2018). Even though BLRaV is suggested to be the main causal agent of BLRD, other viruses could possibly contribute to symptom development in cases of mixed infection with BLRaV and CLRV.

For symptoms rating, most of the birch trees from Berlin which had mixed infection of CLRV and BLRaV exhibited leaf roll symptoms. A synergistic interaction between these two viruses which can modify viral traits such as host range, transmission rates and as well as symptomology cannot be ruled out. Apart from BLRaV/CLRV infection detected in 9 trees (12.5%), there was a mixed infection of BLRaV/ApMV 11.1% (8), CLRV/ApMV 2.8% (2), BLRaV/BiCV 1.4% (1), and CLRV/BiCV 1.4% (1) in symptomatic leaves. The seldom distribution or detection of double infection for these viruses might be due to seasonal fluctuations, different mode of vector transmission as well as synergistic interaction between two viral species and host as reported by (Elena *et al.*, 2014).

The detection of viruses is also limited by the way of sample collection. This because for most of the viruses, it is unknown if they reproduce mainly in leaves or other tissues (e.g. root, phloem, fruits, seeds, flowers, pollen etc.). In this study the focus was on symptomatic leaves. But in further studies it might be of interest to sample other tissues to rule out the tissue of highest virus titre. Those tissues can give also hints on the mode of transmission. In a mixed viral infection, the tissues preferred by the individual viruses might differ in dependence of the virus type and can have its own dynamics within the seasons of the year. Each viral population is influenced by the environment hence, the success of virus replication within the host in a mixed infection depends not only on its adaptation to the host, but also on how its behaviour interacts with that of others within a given environment (Elena *et al.*, 2014).

Synergistic interactions between different viruses and viroids in mixed infection can lead to increased virulence (e.g. symptoms and/or viral accumulation) or even new diseases (Syller, 2012; Moreno and López-Moya, 2020). This is because mixed viral infection causes variety within host virus-vrus interactions which may results in new variants with genetic features. This can cause change in genetic structure of the viral population and could be one of the reasons for prevalence of mixed infection of CLRV and BLRaV infection especially in samples from Berlin. During synergystic interaction, at least one of the viruses' benefits from the presence of the other which are manisfested based on viral titer or pathogenicity or other properties such as the capacity to disseminate efficiently by the vector organism.

Symptoms from mixed viral infection are reported to be often more severe than single infections of the same viruses (Moreno and López-Moya, 2020). This is in contrast with the mixed viral infection examined from Berlin. The symptoms were relatively not severe compared to single infections recorded. Consequently, synergisms may have a high economic impact when they occur in crops because the resulting diseases are likely to become more severe. For instance, the serious epidemic of Cassava mosaic disease, due to mixed infection between isolates of *African cassava mosaic virus* (ACMV) and isolates of the Uganda strain of East *African cassava mosaic virus,* both members of the genus Begomovirus causes huge annual economic losses (Legg and Fauquet 2004). Mixed infections of two viruses also enable recombination, which in addition to mutation, is another source of genetic variation and emergence of new viruses. The existence of more than one virus makes it difficult in understanding the etiology of diseases. It is therefore very crucial to understand the viral pathogenesis as well as evolution as reported by Read and Taylor, 2001 and development of stable control measures.

As there were symptomatic leaf samples without specific record of the four investigated viruses, it is obvious that, there are still unknown viruses are involved in the symptomatology. And in addition it has to be considered that the time of sampling, the storage and different strains of viruses may influence test results.   In 19.4% (14) of birch trees, no specific detection of the four investigated viruses was recorded although some of them exhibited symptoms of chlorosis, ringspot and vein banding (Opoku *et al.*, 2018). This could be due to limited genetic information of viruses in the beginning as well as primer strain specificity. Due to diversity of viruses, some strains are different hence primers used under this study could not detect such viruses. That notwithstanding, there might be unknown viruses or limitation in RT-PCR detection method.

Since the movement of viruses were not equally distributed, movement dynamics of viruses within the trees can differ between viral species resulting in a non-homogenous distribution. A further experiment with newly designed primers confirmed presence of new virus-like sequences which would be discussed (chapter 4.5.5 to chapter 4.5.8).

4.3 Selected leaf symptoms from *Betula* stands -Wildberg/Nagold -Germany

Wildberg/Nagold seed plantation is established for seed production from the phenotypical best trees and clones of (*Betula* sp.) which are further distributed to urban areas for tree planting, landscape and reforestation. Birch leaf symptoms from Wildberg/Nagold were less distinct and obvious compared to those leaf samples from trees of Berlin and Rovaniemi. The trees are artificially stressed for induction of enhanced seed production by crown removal. This is a similar situation in Berlin where street trees are stressed by removal of lower branches for traffic security. The symptomatic leaves were heavily covered by insect bites. The presence of high rates of vectors such as aphids made it difficult to distinguish viral leaf symptoms from insect's bites because both were overlapping in trees. Rating of trees in early summer is an easy way for symptom observation, because the symptoms development is still fresh and not influenced by other factors. These symptoms might change dynamically during the vegetation period in dependence on environmental conditions such as; drought, cutting and grafting, nutrient deficiency as well as high and low temperatures.

Since birch trees do have an active metabolism in terms of protein formation, active cell division, fast and much sugar production and replication (Galibina *et al.*, 2015), viruses can easily benefit during this time and reproduce itself making leaf symptoms visible. These plant viruses' replicates and move between cells via plasmodesmata, symplasmic channels in the adjoining cells walls (Heinlein, 2015). Although the process of virus movement can be complex and requires support by the coordinated activity of several virus-and host-encoded proteins, many viruses achieve their movement with the help of classical, virus-encoded movement proteins (MP) that bind nucleic acids, targets and dilate Plasmodesmata (Lucas, 2006).

During a visual rating of viral symptoms, BLRaV associated symptoms were detected in only two birch trees (E58125 and E58145). HTS and RT-PCR confirmed BLRaV infection of leaves which had symptoms of vein banding, chlorosis and leaf roll. A virus belonging to the genus badnavirus *Cacao mild mosaic virus* (CaMMV) was reported by Puig *et al.*, 2021 to be transmitted by mealybug in Florida.

Since badnaviruses are known to be transmitted by insect vectors, it was a strong argument for the transmission of BLRaV by insects due to high population rate of insects in Wildberg/Nagold. But the two symptomatic trees remained the only infected trees in the plantation so far after investigation by RT-PCR.

Given the high numbers of insects in the seed plantation, a wider spread of the BLRaV was expected if insect transmission is assumed as the main mechanism of virus transmission in the plantation. Furthermore, a mechanical transmission was shown experimentally, which lead to the hypothesis and expectation of a wider spread of the virus BLRaV within the plantation. With only two trees infected by BLRaV in the whole seed plantation, the question arises if the mechanism to transmit the BLRaV is associated with a very specific vector or if another mode of transmission or other dynamics of the transmission exist under natural conditions. Since it is known that BLRaV is mechanically transmissible, cultural practices such as ploughing, pruning tree crowns, cutting branches may lead to possible BLRaV transmission but may be with a low frequency. The thinning of crowns and the seed harvesting practice by cutting whole branches in seed plantations is a common practice whereby the seed production is increase by stressing trees.

The trees in Wildberg/Nagold are regularly stressed by cutting with tools which are not disinfected during the tree management and could one of means for BLRaV transmission. Until now, Badnaviruses which were known to be prominent in tropical areas have now been detected in Europe (Bhat, *et al.*, 2016) especially in Germany and Finland (Rumbou *et al.*, 2018). So far even though some Badnaviruses (e.g *Cacoa swollen shoot virus*) have been found in pollen (Ameyaw *et al.*, 2014), there are possibilities of viruses being harbored in pollen and seeds which could be dispersed by wind or by pollination. BLRaV affecting birches have not been found in pollen (Ingwerson, 2021).

Birch population which has developed under natural conditions are known to cope better with viral diseases because the fittest and viral free survives (Büttner *et al.*, 2013; 2022). The situation is different with the cultivation of seedlings in the nursery. The hypothesis is that a major source of viruses affecting urban trees (e.g. birches) is derived from tree nursery stock grown under inadequate hygienic conditions, without having viral diseases in mind. From the economic point of view, nursery trees are just grown under ideal conditions just to optimize profit (e.g. not to lose a single seedling in pot, nutrients and save labor). This results in the survival of virus infected seedlings which are then transferred to the final stands in the cities. The health status of tree seedlings is comprised at the expense of the labor and cost.

Since seedlings for street trees are not certified as virus- free, unlike the fruit growing trees, there is an increased incidence of viral diseases in urban green space. Due to the little attention given to viruses affecting trees (Nienhaus, 1985; Büttner *et al.*, 2022, Bandte *et al.* 2022), these viruses detected in tree seedling go unnoticed but can be a contributing factor in tree decline.

With the results obtained for the Wildberg-Nagold samples, (i.e. BLRaV 2/71 trees), it can be argued that infection of plant viruses (BLRaV) begins from the mother trees and their seeds or seedlings. For many years, transmission of viruses through seed was considered irrelevant due to fact that seed embryos resist invasion by viruses from surrounding material tissue as reported by Johansen *et al.*, 1994. Again, the presence of virus in the embryo or seed coat is not enough to classify virus as seed-transmitted but instead the virus must remain viable during maturation of the seed and be able to infect seedling after germination. There is however a little attention given to plant viruses infecting tree seedlings, which is presumed to be relevant. Tools used for pruning and thinning out of young seedlings should be disinfected before and after use to help reduce viral infections in tree nurseries by interruption of mechanical transmission.

The management of tree nurseries, such as implementation of certification of tree seedlings for urban environments should be taken into consideration. This would help curb down the rate of spread of viruses and increase quality of seed production.

Seed users have to be ready to bear the cost involved in the high seed quality production and efforts have to be made to develop methods for viral free cultivation and production of trees. Different methods of in-vitro thermotherapy such as shoot tip culture, chemotherapy; micro grafting or shoot tip cryotherapy which are being used in grapevine production (Miljanić, *et al.*, 2022) can also be used in birch seedling plantation. By heating infected cultures at 35°C-42°C for about four to six weeks, some viruses can be eliminated in birch seedlings depending on the type of virus and plant species and then tested using molecular diagnostic techniques.

4.4 Selected *Betula* stands -Rovaniemi -Finland

Suspected viral symptom observation in summer (May 2017), in some areas of Rovaniemi confirmed the presence of not just CLRV but also BLRaV in this study. Based on the observation made during sampling, it can be concluded that, trees from different areas in Rovaniemi appeared weaker with much severe symptoms (Jalkanen *et al.*, 2007) compared to trees from Berlin and Wildberg/Nagold.

Birch trees exhibiting leaf symptoms of viral diseases during summer, 2006 throughout Finland was distinct with CLRV-infected birches (Jalkanen *et al.* 2007). About 42% of the birch trees had symptoms of intercostal chlorosis (weak or severe) which corresponds to what was reported by von Bargen *et al.,* 2012. CLRV was detected in only two out of 26 birch trees from Rovaniemi investigated with the rest having BLRaV infection.

Before HTS brought more insight of the birch virome, CLRV seemed to be the most dominant virus after testing and confirming CLRV- infection via ELISA and PCR. In 2018 first studies on the virome of birches were published by (Rumbou *et al.,* 2018) presenting Birch leaf roll disease (BLRD). *Apple mosaic virus* (ApMV) and *Birch carlavirus* (BiCV) were not detected in all symptomatic leaves investigated from Rovaniemi in 2017, but the presence of BLRaV was confirmed. Jalkanen *et al.,* 2007 observed that during late summer, symptomatic parts of tree canopies could easily be distinguished from asymptomatic parts or closed crowns due to their colour. The entire crown could become symptomatic in few years indicating the susceptibility of most trees to CLRV in Finland. Studies in later years confirmed that birch trees are not only susceptible to CLRV but also BLRaV in Finland (Rumbou *et al.,* 2018).

Viruses do influence host metabolism which is seen by symptom expression. For instance, there are projections that the rapid expansion of human activities and climate change will alter the frequency and duration of virus epidemics (Jones, 2009). Since viruses have diverse transmissible mechanism such as vectors, they could easily be transferred by human activities through trade or vehicular movements. The climate in Finland for instance is gradually becoming warmer due to climate change and it has been projected that global warming associated with increase in annual cropping period, area of cultivation; crop diversity will cause increase in diversity of aphid species, and favor greater spread of aphid-borne potato viruses (Jones *et al.,* 2016). For forestry, the incidence of insect population on field (Rovanniemi) where samples were collected could could be a medium for the spread of viruses such as the BLRaV since Aphids is reported to be most successful vectors for plant viruses including badnaviruses due to array of generic and specific features they possess (Harris and Maramorosch, 2014).

In forestry, changes in wind strength due to climate change in Finland can have adverse effect on forest ability to withstand once they are infected with viruses. Depending on whether it's a mast or non-mast year, birches can produce 0.3 to 0.6 million seeds (Tiebel *et al.*, 2020) and dispersed by wind at an average mean of 40 and 360m (Huth 2009; Tiebel *et al.*, 2020). Hence they can travel hundreds of meters by wind dispersal which could also lead to spread of viruses (such as BLRaV and CLRV) once seeds with viral infections find suitable conditions for growth.

Climatic conditions are known to affect geographic distribution of viral hosts and vectors since the epidemiology of viruses depends on these hosts for propagation (Jones and Barbetti, 2012). In addition to that, the rapid change in temperature during winter (cold stress) can cause damage to trees such as *Olea europaea* as reported by (Petruccelli *et al.*, 2022). With birch also being a tree, they are exposed to viruses which can take advantage of it and replicates when conditions are favorable in summer. Differences in long winter nights and long summer days may also influence the activities of viruses in Fennoscandia. This is because, in winter days, trees have less light to produce biomass especially in the north and can influence the viral replication dynamics. Viruses can influence woody trees (e.g. birches) and indirectly affect the stability of trees and the equilibrium with its microbiome.

4.5 RT-PCR detection of plant viruses

The RNA extraction protocol (Boom *et al.*, 1990) described in this thesis was used in routine detection of a number of plant viruses in woody hosts using reverse transcriptase polymerase chain reaction (RT-PCR) which is rapid and reduces the effect of phenolic compounds and other inhibitory substances. High quality RNA suitable for use in RT-PCR was extracted from birch leaves (from Berlin, Rovaniemi and Wildberg/Nagold) for the detection of BLRaV, CLRV, BiCV and ApMV. During the course of optimizing the RT-PCR protocol for each virus, the reagents from the reaction mixture used for hybridization of the primers at the exact location in the genome available for binding to the desired region was considered. The nad5 transcript in gel electrophoresis provided significant result with specific amplification of product from plant nad5 mRNA as reported by Menzel *et al.*, 2002. There were however some discrepancies in some of the nad5 results and had to be repeated.

4.5.1 *Birch leaf roll associated virus* (BLRaV)

The detection of BLRaV with accession number (MG686419.1; MG686420.1) affecting birches which is known to be a contributing factor to Birch leaf roll disease (BLRD) was found in HTS data in 2014 and first published by Division of phytomedicine-HUB (Rumbou *et al.*, 2018). This was also confirmed in the birches investigated in Berlin, Rovanniemi and Wildberg/Nagold. A sensitive amplification procedure reverse transcription-polymerase chain reaction (RT-PCR) used for detection of plant viruses with RNA genomes was reported by (Henson and French, 1993). After successful generation of primers for RT- PCR detection, different strains of BLRaV have been detected with optimized primer sets which were derived from BLRaV strains affecting birches from Berlin and Corsica.

Based on conserved region from BLRaV, the primers were successfully validated and published by (Pack *et al.*, 2019). The use of primers (BadnaK-FW/ BadnaK-RV, forward and reverse primers) developed from Corsica strain and (Badna SG-FW/BadnaSG-RV) from *Schwarzer Grund* (Berlin, Germany) produced specific PCR results investigated in this study without any side reaction (primer dimers) for birch leave samples which was also confirmed by Pack *et al.*, 2019. Nevertheless, it is unknown if those primers can also detect all Scandinavian strains. Rumbou *et al.*, 2018 reported that, in a single birch tree, there are several different strains. Since the primers were developed from sequence information from different sites, it is reasonable to assume that the primer pair BadnaK-FW/BadnaK-RV binds more specifically to a cDNA template from a sample from Corsica than to a sequence from another site. They could also bind specifically to the cDNA template from Rovaniemi as well as Berlin. There was no doubt, the primer pair BadnaSG-FW/BadnaSG-RV chosen from the HTS sequence data of a leaf sample from Schwarzer Grund binded specifically to the virus sequence of the virus variant from Berlin-Germany.

Both primer sets were quite sensitive and hybridized well at the coat protein coding region of the BLRaV genome with a given PCR protocols without any side reactions such as primer dimers or non-specific products especially for samples from Berlin. The primers used for RT-PCR was quite robust, allowing detection of BLRaV in (40) of the (72) birch samples from Berlin, (23) from (26) birch samples from Rovaniemi and (2) from total of 71 birches from Wildberg/Nagold seed plantation. The difference in the ease of detecting BLRaV by RT-PCR may reflect differences between plant species and virus titers, different cultivation sites or due to quality and quantity of transcribed RNA extract from birch leaf samples. Area of binding sites is conserved in the coat protein without any wobbles in the nucleotide sequence alignment as shown in (Appendix 11.3; Tab. A1) after BLAST.

The reason for no detection of BLRaV in some birch leaves with the BLRaV primers suggests that there is some sequence variability among viruses, although the extent of the variation was not determined. This problem of sequence variability might be overcome by combining primers in a multiplex PCR assay. Alternatively, more conserved regions of the genome could be targeted for design of the primers. There were sometimes non-specificities, making it difficult to correctly evaluate the results. If the primers bind non-specifically, this lead to formation of unwanted by-products due to competition in the PCR reaction.

The reagents from the extraction mixture are then used up for hybridization of primers at an incorrect region in the genome making it unavailable for binding to desired region (Mülhardt, 2010) hence producing a lower yield of PCR product. To prevent non-specific binding, there is a need to adjust the PCR components such as the Mg^{2+} concentration (Lorenz *et al.*, 2012) or using another diagnostic tool such as the real time polymerase chain reaction (real time-PCR). The sensitivity of PCR also allowed detection of BLRaV in *Chenopodium quinoa* after successful mechanical inoculation of BLRaV from birch leaf to host plant. This result confirms earlier data, based on mechanical transmission experiments that showed this species to be a host of the virus (Bhat *et al.*, 2016). The incidence of BLRaV was very prevalent in all samples from in Berlin, Rovaniemi and but not in Wildberg/Nagold. With the diversity of symptoms observed, correlation of symptoms to viruses was very difficult. However, symptoms of leaf roll, chlorosis and vein banding were associated with BLRaV.

HTS results from two *Betula* tree samples (E58128 and E58145) with symptoms leaf roll; chlorosis and vein banding were confirmed to be infected with BLRaV. This was then tested by RT-PCR as well as Sanger sequencing to prove the validity of BLRaV. An identity matrix from contig number (*H0515_contig41 and H0515_contig 40*) gave percentage identity of 92% and 94% respectively indicating how closed they are related. This corresponds to the report by Rumbou *et al.*, 2018 where symptoms of chlorosis and vein bending was associated to BLRaV. It was obvious to detect BLRaV in Wildberg/Nagold due to the presence of insects on the field. Although we could not investigate insects (e.g. aphids), Badnaviruses are known to be vector transmitters as reported by Bhat *et al.*, 2016.

Since leaf samples were collected from a seed plantation (Wildberg/Nagold), cultural practices such as pruning, thinning out were done with equipment such as scissors and secateurs. This equipment when not disinfected after use could lead to spread of BLRaV. There is much need to focus on trees in the nursery because infection of viruses could begin from the field.

Having found bird nest on birches as habitat for birds and with the knowledge of birds feeding from birch seeds, it is unclear whether they could also be a means of transfer of BLRaV from one location to the another like the island Corsica. With our findings and results, we could argue that infection of BLRaV could begin from field of plantation before being transported to the streets or urban green space. This confirms to what was reported by Büttner *et al*., 2013; 2022 suggesting viruses appear to be more widespread than previously thought. They are prevalent in urban green space especially in Berlin.

Badnaviruses are known to be distributed in tropical and temperate regions and infects tropical crops such as cocoa, citrus, taro, black pepper and yam have recently been detected in Europe. The first badnavirus to be reported in *Betula* species is the BLRaV (Rumbou *et al*., 2018). After the first detection, BLRaV have been detected in Germany (Berlin), France (Corsica), Sweden, and Norway. This is of great importance not only to increase the list of hosts affected by members of this genus but also to expand the genus geographic distribution towards its northern limit. It is however interesting how BLRaV are being detected in tree grown in temperate regions. With knowledge of Badnaviruses often found in tropical host plant, it can now be disputed that they also exist in temperate climatic regions with the birch as a host. The result obtained confirms the random distribution of BLRaV along the streets of Berlin. The samples collected from *Tempelhofer Feld* were dispersed and seems not be cultivated with less symptoms compared to what was observed in other areas.

BLRaV was detected in (3) out of 5 tree samples even though they seem not to have been cultivated. Although some streets had infection of other viruses, BLRaV was prominent in all areas. This could be as results of climate change, international trade, and movement of vectors and shipment of seedlings as reported by Rubio *et al*., 2020. There is report that the economic loss caused by the different species of badnavirus in various crops varies between 10% and 90% (Bhat *et al*., 2016) which is an indication that BLRaV can also have economic impact on *Betula* species.

The viral infection on woody plant may take a long time for them to be damaged. This is because *Betula* trees which are a deciduous can harbor or absorb as many complex viruses and adapt to metabolic mechanisms of tree. For example, a natural growing birch tree can survive better than trees which are cultivated for planting. Having said that, a long term effects on trees could be detrimental since reports suggest that Badnavirus can cause huge impact (i.e. yield quality) on crops such as Cocoa (Ameyaw *et al*., 2014). Posnette, 1940 are also reported that the decline of cocoa tree results to death (3 to 5years) after symptoms development such vein chlorosis, vein banding, stem and root swelling and pod deformations.

The symptoms leading to death of cocoa trees are similar to what was observed in birch leaves; hence BLRaV may lead to death of birches after some years. On the other hand, due to the ability of birches to withstand environmental stress as well as harbor viruses, they can resist or find a mechanism to disrupt viruses from replicating. The mixed infection of BLRaV and CLRV which was dominant in most of the investigated trees suggest that, there could be a synergistic interaction between these viruses which was formally attributed to only CLRV causing Birch leaf roll disease (Rumbou *et al.*, 2018).

The observation made on symptoms caused by BLRaV and CLRV were similar with symptoms of chlorosis and leaf roll being rampant especially in leaf samples from Berlin.

Based on the results from BLRaV, we could argue that, infection of BLRaV begins from field since we could detect BLRaV in symptomatic birch leaves from Wildberg/Nagold plantation. Seed production trees were confirmed to be infected by BLRaV. There is therefore the need to give much attention to plant viruses especially the BLRaV in urban tree management. Management practices such as pruning, thinning out, grafting in the nursery should be carefully done with the right tool and disinfected before and after use. Proper and more reliable diagnostic tools such as Rolling circle amplification; reverse transcription polymerase chain reaction as well as others should be improvised and used for efficient detection of BLRaV. There should be the need for awareness through media, journals, etc. that viruses exist in urban green space hence a need to find ways of reducing the spread.

4.5.2 *Cherry leaf roll virus* (CLRV)

The design of primers for the detection on the genus or species level is always necessary if a species consist of a high diversity of strains which cannot be detected by a single primer set. In case of CLRV, that would mean that all strains considered belonging to CLRV are covered by those detection systems or on the genus level that all species belonging to the genus *Nepovirus* are detected by one primer pair. The question arises if primers designed for CLRV covers all strains and are also disguisable by serological methods.

The primer pair combination (CP350F/CP977R and CP188F/CP350R) used for detection of CLRV in our investigated samples resulted in amplification of coat protein region of CLRV which is in confirmation of what was reported by von Bargen *et al.*, 2012; Langer *et al.*, 2016. CLRV was however detected in (2) birch trees out of 26 in Rovaniemi-Finland. For Berlin samples, the primer pair (CP350F/CP977R and CP188F/CP350R) were strain specific and could be detected in (19) birches out of 72 trees.

The two primer pair was robust and PCR samples were amplified at the expected size for samples from Berlin but not for Wildberg/Nagold samples. They produced unspecific bands which suggest the primers could not target the coat protein regions. There are reports of several strains for CLRV such as *Malus domestica, Ribes rubrum, Rumex obtusifolius, Vaccinium darrowii,* (Woo and Pearson, 2014) and *Juglans regia* L. (Cieślińska, 2020). There are possibilities that trees from Wildberg/Nagold were infected with CLRV but these primers could not detect them. CLRV was not detected in samples from Wildberg (Nagold), which seems to be unlikely, because symptoms of vein banding and leaf roll normally associated with CLRV were observed in two particular trees (E58128 and E58145) and was suspected to be infected by CLRV.

The primer pair CP350F/CP977R and CP188F/CP350R did not amplify any fragment, although amplicons could be generated in leaf samples with the primer combination CP188F/CP350R for Berlin samples. The virus variants of the CLRV in birch seem to show a high degree of variability and can only be detected using a primer pair that can specifically bind to the present sequence. Based on information from the (NCBI), most primers have been detected in conserved regions such as the coat protein and untranslated region (UTR).

A bioinformatical evaluation of the primers hybridizing in the coat protein area by blast via NCBI on well-known primers (CLRV-CP350F/ CLRV-CP977 R) used for CLRV detection resulted in wobbles indicating unconserved nature of the particular primer binding region. For the forward primer (CLRV-CP350F), an accession number (S63537.1) and (MK402282.1) were deduced and seems to include wobbles among the nucleotides. These could the reason why primers could not work (additional products in PCR) and sometimes produces primer dimers. Based on that, reverse primer of the primer set targeting the 3' UTR (RW1/RW2; Werner *et al.*, 1997) was modified by Division of Phytomedicine, HUB and published as RW2modified by Bütow, 2013. These new primer pairs produced by Bütow *et al.*, 2013 are sensitive and produce specific bands but were however not used in our investigations.

CLRV is transmitted naturally by seed and pollen (Card *et al.*, 2007) and that should in further investigation be taken as a target tissue for detection of CLRV in addition to leaves. The insect vector aphid (*Kleidocerys resedae*) has also been reported by Bandte *et al.*, 2011; Langer *et al.*, 2013 as means of CLRV.

Having found insect in most of the Berlin sites especially in Schlangenbader straße (Charlottenburg Wilmersdorf) and Rabindranath-Tagore- Straße (treptow-Köpenick), it can be argued that presence of *K. resedae* could be a potential vector for CLRV distribution eventhough insects were not tested for CLRV as described by (IC)-RT-PCR. Most of the trees with CLRV infection in Berlin were found in younger trees (about 10 to 15 years) in Gleiesdreieck, Mauerpark, and Blankensteinpark. The older trees (about 80 years) with CLRV were found in Schlangenbader Strasse and Grünau.

As CLRV was found in younger and older trees, one can argue that, birch trees can combat the infection for a long time under specific conditions and CLRV in a non-stressed tree will not seriously affect the tree. Whether the longevity or tree height, trunk diameter or other growth parameters of the trees are influenced by CLRV is still unknown. CLRV has a significant economic impact in other hosts and has been reported to cause leaf rolling and death of cherry plant in England, death of walnut trees in Hungary with an estimated crop loss of 91-98% in Germany (Woo and Pearson, 2014) and decline of birch forest (Büttner *et al.*, 2013).

The rate of increase of CLRV population birches in Berlin could have adverse effect on birch population such as tree damage which could cause decline of these pioneer trees as well as their quality. For example, the CLRV isolates affecting Walnut have been reported to cause economic loss in Walnut production (Cieślińska, 2020). Since CLRV have been detected in birches, there could be an economic impact or loss on them hence need for prophylactic measures and good management practices.

4.5.3 *Birch Carlavirus* (BiCV)

Birch Carlavirus were detected in (6) out of (72) tree samples from Berlin but not Wildberg/nagold and Rovaniemi. For the detection of carlaviruses in birch, the primer pair CarlaK-FW/CarlaK-RV in the area of the polymerase was chosen, since suitable sequence information was only available for this. The contig "1008" (Corsica) has 61.3% similarity to the genome of the *Elderberry carlavirus D isolate* (KJ572563.2) and 57.8% to the genome of the *Carlaviruses Helleborus net necrosis virus* (NC_012038.1). These primers set combination (Carla K-FW and Carla K-RW) confirmed the presence of BiCV in birches from Berlin after gel electrophoresis at an amplicon size of 197 bp from coat protein transcript (RT-PCR). The reason for no detection of BiCV from Wildberg/Nagiold and Rovanniemi could be attributed to primers not being specific.

The RT-PCR systems for detection of BiCV in birch could be rather strain specific than genus specific. Since primers were designed from strains from Corsica, the probability of binding to specific sites was not that significant. A total of 53 genomes for carlaviruses and 70 badnavirus genomes are stored in the NCBI database (as of June 2022). The heterogeneity within these two groups is very high (Borah *et al.* 2013; Chen *et al.* 2006). So far, the full-length genomic sequence of this novel virus has been deposited at GenBank under accession number MH536506 (Birch carlavirus isolate BpenGer407526-5M) (Rumbou *et al.*, 2020). Recent comparison of these primers with sequences from NCBI has shown that, the sequence of BiCV primers seem not to be conserved and it indicates less similarity at the 3' ends. Applying BLAST for the creation of an alignment, there were presence of wobbles in the primer binding sites (MH536506.1) for Carla K-FW forward and MH536506.1 for Carla K-RW) (Appendix 11.3, Tab A1) and could be a reason for no detection in Wildberg/Nagold and Rovanniemi samples.

There is still a need for further evaluation and optimization of these primers as they are strain specific which is currently being done (Köpke *et al.*, 2020). Leaf samples infected with BiCV had symptoms of intercostal chlorosis and necrosis. Although these symptoms have been associated with *Helleborus spp*, (Eastwell *et al.*, 2009), much research still needs to be done to confirm the symptom correlations to the carlaviruses. Symptomatology may differ in different parts of the canopy and could presumably be related different virus populations in birches or to other parameter (Rumbou *et al.*, 2016). Due to the high variability of the virus sequences found, it is reasonable to assume that there are regionally specific virus variants within the wide distribution range of the viruses. A mixed infection of BiCV and BLRaV was confirmed by RT-PCR.

Since it has been confirmed that some carlaviruses interact with other viruses and causes serious diseases (e.g. symptomless *lily mottle virus* interacts with *cucumber mosaic virus* to causing fleck disease (Niimi *et al.*, 2003), the mixed infection of both viruses (BiCV and BLRaV) could also cause serious disease in birch. Their mode of transmission has been reported to be via vegetative propagation and primarily by aphids, whiteflies and occasionally by seeds. The place of main replication and highest titre (specific tissue) of BiCV in birches is still unknown, but based on the above-mentioned way of transmission known from other carlaviruses the investigation of leaves seems to make sense.

4.5.4 *Apple mosaic virus* (ApMV)

Until now little information is known about ApMV affecting birches although some positive results have been detected by RT-PCR and ELISA since a long time (Grüntzig *et al.*, 1996; Bandte *et al.*, 2010; Büttner *et al.*, 2022). With a wide variety of host range, ApMV is known to affect *Betula* spp. (*Betula pendula, B. papyrifiera, B. alleghaniensis*), European mountain ash (*Sorbus aucuparia*), Horse chestnut (*Aesculus hippocastanum*), raspberry, hop (*Humulus lupulus*), European elder, mahaleb cherry (*Prunus mahaleb*) and several species of *Trebouxia lichen* (Fulton, 1972; Kanno *et al.*, 1993; Grimová *et al.*, 2013; Petrzik *et al.* 2014). A BLASTn contig annotation of the ApMV primer sequences applied in this study was done via (NCBI) to check if they are found in conserved regions without wobbles.

In some cases, the primers could cross hybridize with unknown parts in the plant genome or microbiological background of different natural samples leading to bigger nonspecific products which is evaluated as negative. The partial amplification of ApMV-coat protein in RT-PCR from the leaf samples with the primer combination CP153F/CP1739R reported by (Langer, 2016) was confirmed in this investigation. For mixed infection, there was record on triple mixed infected (ApMV/CLRV/BiCV) individual birch trees from Berlin samples but it is unknown if they have influence on the symptomatology or interplay together in a synergistic way.

Even though ApMV were seldomly found in city of Berlin, they might also contribute to decline of birches as economic losses are known for this virus. The presence of apple viruses in apple trees causes a yield reduction of about 9-46% (Cembali *et al.*, 2003). With the effect of ApMV on fruits trees, it can be argued that ApMV can also cause reduction in quality of birch seed plantation or wood production for a long time. ApMV which belongs to group of Ilarviruses are mechanically transmitted by insect thrips feeding on pollen grains containing the virus or by pollen grains contaminated by the virus (ICTV report, 2022).

In contrast to that, ApMV has not yet been detected in birch pollen. This might be as a result of a low percentage of ApMV-infected embryos or because the pollen is too tiny to produce results (Digiaro and Savino., 1992). Having said that, environmental stress factors such as temperature and drought cannot be ruled out.

4.5.5 Partial Benyvirus-like signature

Two contigs (01516 and 01856) with similarity to Benyviruses were found in HTS samples from 2015 (Finland in tree Bpub20_15396 and dsRNA BetulapubescensFinn-Bpub20 with HTS number H008 and H0204).

A 480 bp contig 01856 from the birch HTS data showed 28% identity to the Benyvirus *Rice stripe necrosis virus* (RNA1 6640 bp) replication associated protein and a second contig01516 (539bp) showed 43% identity to the RdRP of *Beet necrotic yellow vein virus* (RNA1 6750 bp). By Sanger sequencing the existence of benyvirus-like sequence in the original tree Bpub20 from Rovaniemi (Finland) was confirmed for the primer Beny 1856FW but was not done for the primer Beny 1856RW (overlapping of non-specific sequences). The sequencing further confirmed the presence of viral signature in the original leaf sample. The optimization of Benyvirus-like PCR did not give a specific amplification product from other trees.

The RNA1 was artificially assembled by Phytomedicine-HUB with a coverage of 6082 bp using contigs H008, *Bet_Finland_2014c_dsRNA (dsRNABetulapubescensFin-Bpub20)* and H02303 Bet_Finland_2016a (6-Bpub20_15396). Since Benyviruses are found in sugar beet, rice and burdock, this could be a first step in recognition of Benyvirus-like sequence in woody host. Once the complete genome of the Benyvirus has been confirmed in birches, future planting of birch seedlings would need to take into account quantifying and detecting viruses directly in the soil, monitoring the spread of viruliferous plasmodiophorids, and effective disease management strategies as reported by (Rush, 2003).

With a strict host range, they are transmitted by root-infecting vectors in the *Plasmodiphorales* family which was initially described as fungi but now classified as Cercozoa (*Protozoa*) (Gilmer *et al.*, 2017). With two genome segments of polyadenylated positive-sense RNA (approximately 4.6 and 7kb and up to three additional RNA components of 1.3-1. 8kb), the *Beet necrotic yellow vein virus* (BNYVV) has been reported to be widespread in Europe, North America and Asia and causes severe damage to soil borne *rhizomania* disease of sugar beet (Gilmer *et al.*, 2017). The PCR optimization of Benyvirus-like sequence did not give specific amplification product in other trees and this may be due to poor plant materials stored for longer time.

There were also primer dimers which could be attributed to non-optimized composition of PCR such as $MgCl_2$ which is noted for causing most of primer dimers and undesired PCR products when not correctly pippeted (Lorenz, 2012). Other birch tissues than leaves were not investigated although an investigation of root seem to be of interest in further investigation.

4.5.6 Birch Capillovirus-like sequence

Partial amplification of capillovirus-like sequence at an expected size of 469bp was detected in most of *Betula* samples from Berlin but not always from Wilberg/Nagold. However, 5 out of 10 samples from Berlin tested by RT-PCR in 2018 confirmed the presence of only Capillovirus-like in birches although the amplification of nad5 transcript showed no inhibition PCR. For Wildberg samples, 56 capillovirus-like signatures were detected and confirmed by Sanger sequencing (Appendix 11.4: Fig. A3). The partial sequence of a Birch capillovirus-like sequence (MK402234.1 and MK402233.1) was published by Division of phytomedicine (Rumbou *et al.*, 2020). First hints on this viral signature came from HTS from trees in Finland (BpubFinn407501-3A) and Germany (BpenGer407526-B5) with some visible symptoms.

The sequence 821bp long contig has significant similarity to *Citrus tatter virus* coat protein (e-value 2e-20, 28, 45% identity) or *Yacon virus* A polyprotein (E-value 2e-19, 29 % identity). Later, the birch HTS data confirmed the region in several other HTS data (2016, 2019 and 2020). The existence of a polyA sequence in the terminal region provided the idea of the existence of a true viral signature. Capilloviruses are positively-oriented and with a linear single-stranded RNA genome with a size of 6.5-7.5kb (Adams *et al.*, 2012) exhibit a restricted host range. *Apple stem grooving virus* (ASGV) is reported to naturally infect some perennial plant species that are not grafted, such as lily (*Lilium longiflorum*), wild fig (*Ficus palmata*), and Himalayan raspberry (*Rubus ellipticus*) (Bhardwaj and Hallan 2019; Inouye *et al.* 1979).

Mechanical transmission of ASGV to *Chenopodium quinoa* and *Nicotiana benthamiana* plants has also been experimentally achieved (Clover *et al.* 2003; Inouye *et al.* 1979). However, woody plants are generally difficult to infect with plant viruses via mechanical means (Yamagishi *et al.*, 2010). This may be due to sap transmission of plant virus, the inoculum source, buffer composition and growth environment after inoculation. No vector transmission is known for ASGV (Inouye *et al.* 1979) and *Cherry virus A* (CVA), but transmission of *Citrus tatter leaf virus* (CTLV) via seeds has been reported in Citrus seedlings (Tanner *et al.*, 2011). With woody birches, capilloviruses are capable of infecting them. However, no investigation has been conducted on possible mechanical transmission from woody host to herbaceous plants yet.

The Capillovirus-like sequence confirmed by RT-PCR in 56 birch trees from Wildberg-Nagold had less severe symptoms. These results were confirmed by Sanger sequencing indicating the presence of partial capillovirus in birch samples from Wilberg/Nagold and Berlin. Capillovirus-like sequence were prevalent in most of Berlin samples investigated, however, for samples from Wildberg/Nagold it was not detected in all samples but in a few even though amplification of the nad5transcript by RT-PCR showed no inhibition indicating an intact and good RNA quality. This could be a matter of primer specificity as the sequence could easily be amplified from Berlin but not from Wildberg/Nagold (different strains). Due to presence of high rate of insects at Wildberg-Nagold, there could be vector transmission of viruses although the vector transmission of Capillovirus is unknown (Inouye *et al.* 1979).

In the genome of *Betula pendula* (contig 934) with accession number FXXK01000935.1, the Capillovirus-like sequence seems to be integrated as reported by Marais *et al.*, 2018 for Japanese Apricot (*Prunus mume*) with the novel name Mume virus A. A complete genome organization of this virus showed a typical capilloviruses with two overlapping open frames encoding alarge replication associated protein fused to coat proten and a putative movement protein and closely related to *Cherry virus A*(CVA) and *Currant virus A* (CuVA). The Blast via NCBI confirmed the presence of Birch capillovirus with accession number of MK402233.1 and MK402234.1 respectively. This confirms to what was reported by Rumbou *et al.*, 2020 and has been given a tentative name Birch capillovirus. Upstream and downstream of other ORFs proved futile. This gives a first impression of possible integration of Capilloviruses in the *Betula* genome. A hit on the protein level resulted in a poly protein on the ORF1 with a significant E-value of 1e-98.

From this limited information it has to be taken into consideration that this Capillovirus-like sequence belongs to the group of endogenous viral elements representing only parts of an ancestral or interrupted viral integrated sequence. An integration long time ago would also explain why only some trees from Wildberg/Nagold did not contain the sequence. This cannot be confirmed yet but reflects a hypothesis that capilloviruses could be integrated in birch genome. There is therefore a need to find more sequence information for successful completion of the genome and to answer the question if this Capillovirus-like sequence belongs to a true active virus.

4.5.7 Birch Caulimovirus-like sequence

The results from Berlin and Wildberg-Nagold confirmed the presence of partial caulimovirus-like sequence as reported by Rumbou *et al.*, 2020. Out of the (71) investigated trees from Wildberg-Nagold; caulimovirus-like sequence was detected in (69) birch trees. Experiments herein focussed only on contig410 with the goal of a first confirmation in nature and to provide a first look into the distribution of the sequences similar to Caulimoviruses by screening of natural birch trees from Berlin and Wildberg-Nagold in the year 2018 and 2019 respectively.

Based on HTS data and bioinformatics analysis from 2018 originating from *Betula pendula* (Berlin), additional sequences belonging to the group of Caulimoviridae which differ from the BLRaV were identified. One contig 0108 (*2018 Bet_E56736_3 Berlin Rudow 1919 bp - H008*) had significant similarity to a *Dahlia mosaic virus* (gbAGT41978.1 blastx: E value: 3e-28 percentage similarity 61.39%) reverse transcriptase.

Two other contigs (contig410 from 2014 tree in Finland BetulapubescensFin407501-3A; contig1079 from 2014 tree in Finland BetulapubescensFin407501_3A) representing partial sequences similar to the polyprotein of *Dahlia mosaic virus* (contig_410 Blastx: E-value: 6e-24, percent identity: 50.88%, YP_006732334.1) and a putative enzymatic polyprotein of *Rudbeckia flower distortion virus* (contig_ 1079, Blastx: E-value: 2e-46, percent identity: 56.49% YP_002519387.1) were also produced. Both RT-PCR and Sanger sequencing confirmed the presence of partial Caulimovirus-like sequence. A further bioinformatical sequence analysis and database request was conducted in this work.

Due to ability of caulimoviruses to retro-transcribe, and integrate into plant geome (Hull, 2001; Pooggin *et al.*, 2018), the partial detection of caulimovirus-like in birches could also retro transcribe and integrate in birch genome. The information on caulimovirus-like sequence in the birch from HTS might represent several viruses or strains or even endogenous retroviruses (ERVs) as RNA sourced from different original trees. The partial detection of Caulimovirus-like sequence indicates that they are either present or integrated into the *Betula* genome. The *Betula pendula* (contig 62091; accession number CAOK01062084.1) were found to be integrated with Reverse transcriptase (*Aristotelia chilensis virus 1*) polyprotein (*Dahlia mosaic virus D10*) after BLASTx on the protein level (Fig.35). As to whether these integrated viruses are transposons or not remain unknown hence a need for further investigations. Transposable elements (TEs) are able to move from one locus to another in a given genome and to replicate themselves in the process (Bennetzen *et al.*, 2014).

Again they could be ancestral viruses which were accidentally integrated into transposons area in the genome and later disrupted by transposons movement. There seems to be endogenous viral element in the genome *Betula* without ruling out transposable element though Rumbou *et al.*, 2018 reported of nonexistence of it in the BLRaV sequence. Geering *et al.*, 2014 reported that the role of endogenous caulimoviridae sequences to plant genome and its beneficial impacts (if any) on plant are poorly understood, although a role in defending against infection by cognate (same or similar) exogenous virus is a popular hypothesis. If they are integrated they might belong as well to the plant defence system. With well-designed primers, molecular diagnostic tool RT-PCR has given hints of presence of partial birch caulimovirus-like sequence which is still under investigation.

Almost 50% of pathogens causing emerging plant diseases are viruses (Anderson *et al.*, 2004) and both, wild and cultivated plants are vulnerable to multiple infections of different pathogenic viruses (Moreno *et al.*, 2020) or different strains or genotypes of the same virus.

However, some viruses have been shown to be neutral or beneficial to their host plants, providing mutualistic interactions and strengthening the resistance of plant against abiotic stress (Roossinck, 2015; Fraile and Garcı́a-Arenal, 2016; Lefeuvre *et al.*, 2019). These non-pathogenic viruses may often go unnoticed among pathogenic viruses. However, mixed infections combining pathogenic and non-pathogenic viruses have been neglected in plant virus disease research.

As a result, there is lack of understanding as to what extent these virus interactions within the same host could affect the evolutionary dynamics of viral populations (Schoener, 2011). There are reports that, during host-viral interaction, the host especially (in fruit plant) can trigger antiviral response including; RNA interference (RNAi), systemic acquired resistance, hypersensitive response and DNA methylation (Singh *et al.*, 2019).

Small RNAs (sRNA) which are generated from the cleavage of double-stranded RNAs (dsRNAs) (originating from invading viruses) or RNAs with hairpin structures by Dicer-like proteins (DCLs) are loaded onto Argonaute (AGO) protein complexes. This induces gene silencing of their complementary targets by promoting messenger RNA (mRNA) cleavage or degradation, translation inhibition, DNA methylation, and/or histone modifications (Singh *et al.*, 2019). The regulation of RNA activity is referred to as RNA interference (RNAi) which is evolutionarily conserved process in eukaryotes. The mechanism used by plant RNAi plays a significant role in plant immunity against viruses and controlling diseases via genetic engineering. Hence the needs to further develop and implement the use of RNAi to resist economically important viruses from infecting variety of crops and birch trees.

4.5.8 Birch Deltapartitivirus-like sequence

Partial detection of deltapartitivirus-like sequence were confirmed in samples from Berlin and Wildberg-Nagold by RT-PCR but could not be confirmed by Sanger sequencing. The HTS data from a birch in Berlin (2018) (Rudow E56736_3 a contig0324) was identified showing partially similarities to the RNA-dependent RNA polymerase gene (RdRp) belonging to RNA1 of *Lysoka partitivirus* with a size of 434 bp (blastx: E-value: 3e-127 percent identity: 71.01%; AWV67015.1). With the designed primers provided by Division of phytomedicine-HUB, the sequence was confirmed in the original tree (E56736_3) by Sanger sequencing of the 579 bp RT-PCR product.

Deltapartitiviruses are double-stranded RNA (dsRNA) viruses with a bipartite genome. RNA1 was identified in birch, but RNA2 was not yet extracted from the HTS dataset. Each RNA1 and RNA2 consists of one single ORF. RNA1 contains the information for the RdRP and RNA2 for the coat protein. Since there has been a report on *Fig cryptic virus* infecting Fig trees targeting ORF1 and ORF2 via HTS (Chirkov *et al.*, 2021), there is a possibility of this virus integrated in the *Betula* genome. Their mode of transmission has been reported to be by ovule, pollen and seed embryo and is known to induce persistent infections in plants.

In this study only birch leaf samples were investigated and would further investigations should focus on other parts of birches (i.e. pollen and seeds). Although a PCR product was amplified at the expected size of 324bp, Sanger sequencing of samples from Wilberg-Nagold could not confirm the existence of Deltapartitivirus in the investigated leaves.

4.6 Bioassay

Bioassay confirmed experimentally the mechanical transmission of BLRaV directly from symptomatic birch leaves to the host plant *Chenopodium quinoa* (Fig. 16). After 21 days' post inoculation, symptoms of chlorosis and ringspots were observed only in matured host plant *Chenopodium quinoa*. Out of the 20 inoculated biotest plants, only two *Chenopodium quinoa* showed symtpoms of chlorosis, leaf roll and were associated to BLRaV by RT-PCR. This was also confirmed in samples from Wildberg-Nagold where leaf roll and chlorosis were associated to BLRaV as reported by Rumbou *et al.*, 2018. Bioassays confirmed *C. quinoa* and *C. amaranticolor* to be a suitable test plant for studying BLRaV infection. The transfer of virus from BLRaV infected birch leaf to host plants (*C. quinoa and C. amaranticolor*) resulted in development of distinct symptoms such as ringspots and interveinal chlorosis.

The other host plants (*Nicotiana benthamiana, Nicotiana tabacum L, var. samsun*) showed no symptoms and was not infected with BLRaV.This could be that, these bioassay plants are not suitable for BLRaV.

Other factors such as temperature in green house, contaminations, substrate used as well as growing conditions cannot be ruled out. Further evaluation of host plants by RT-PCR confirmed BLRaV transmission only to *Chenopodium quinoa* but not the other host plants. In contrast to that, Rumbou *et al.*, 2009 confimred *Arabidopsis thaliana* as a suitable host plant for CLRV transmission but this host plant was howver not used in our investigation. With a wide host range of Badnaviruses, mechanical inoculation as a means of transmission of Badnaviruses from herbaceous plant to another host plants have been reported by Bhat *et al.*, 2016. This was confirmed in this thesis but this time directly from woody plant to herbaceous plant. However, Breuhahn *et al.*, 2013 confirmed CLRV transmission from Germany and Finnish birches via grafting.

The successful transfer of BLRaV to host plants gives room for further characterization of the BLRaV. Antibodies can be produced once BLRaV has been isolated. Serological methods known to be easy and efficient can then be used for detection of BLRaV once production of antibodies is successful.

The BLRaV can be characterized independently and enriched for viral particle isolation for further investigation by electron microscopy or even production of antibodies for serology or electron microscopy characterization. Isolation of BLRaV and production of antibody is however time consuming and labour intensive and requires a sophisticated technique. A mixed infection (CLRV/BLRaV) was found in one of the host plant (E54095). This gives an idea whether CLRV and BLRaV interact with each other or dependant on each other in a synergistic manner as explained in Chapter 4.2.1.

Since BLRaV has been mechanically transmitted, there is no doubt seedlings in nursery could be infected when tools used in pruning and thinning out are not disinfected before and after use. But the natural spread of the virus seems to follow special mechanism otherwise the spread in the seed plantation of Wildberg/Nagold would have been even worse as the tool used for managing the trees are not disinfected.

Little can be learned directly on the nature of the virus itself until it can be studied apart from its host. There is the need to conduct further research on the mechanism underlying the mechanical transmission of this BLRaV in nature. For instance, it is necessary to determine the developmental stage of the virus before transmission (full developed particles, free dsDNA, and dsRNA or integrated into the host genome).

There could be a possibility of badnavirus transmission via seed but this was not investigated. There was however some difficulty in bioassay experiment such as inoculating the host plants with inoculum. The holes made in the leaves were carefully done to avoid injury to host plants. Once host plants are injured, the virus might not be able to infect the bioassay plant since cells die before they can multiply. The use of celite and inoculation buffer must be carefully mixed well before use.

Despite the challenges, mechanical transmission experiments for BLRaV from infected *Betula* leaf species using bioassay plants (*Chenopodium quinoa*) proved to be effective. Consequences for hygienic practices during tree management and maintenance will be examined once data confirms pathogenicity of Badnavirus strains found in *Betula* spp.

With our findings from Wildberg/Nagold and successful transmission of BLRaV mechanically, it can be argued that infection of BLRaV begins from seedlings but this needs more investigations for several years.

4.7 Serological methods (DAS- ELISA)

To confirm the presence of CLRV affecting leaf material from birch which was forwarded to bioassay experiments, antibodies generated by Phytomedicine-HUB were applied. Viruses can be detected by DAS-ELISA with the help of antibodies. For example, the provision of antibodies (BIOREBA CLRV-e/BIOREBA CLRV-ch) by BIOREBA made it possible to detect CLRV. The main objective of the investigation was to determine if symptomatic leaves which had CLRV infection by RT-PCR could be confirmed by a second method DAS-ELISA. Using the same leaf samples (TT SEAL 0407510 E-54094) which had CLRV infection, the serotype of CLRV (from cherry and elderberry strain) could be addressed.

Both CLRV serotype used were polyclonal and do not recognise all isolates but were complementary to each other. In other words, CLRV-ch isolate reacts with cherry isolate but not isolates in elderberry whiles CLRV-e reacts to elderberry isolates but does not recognise cherry isolates (Clark *et al.*, 1977). The question would arise if those ELISA can detect BLRaV as it was not known at the time of antibody production. There would be a possibility to detect BLRaV once antibodies are produced. For this purpose, 11 symptomatic leaves taken from different parts of a specific tree (TT SEAL 0407510; E54094) were used for the DAS-ELISA test. High serological variability was revealed using the different polyclonal antibodies produced against a cherry isolate of CLRV.

Both reagents (i.e. CLRV-ch and CLRV-e) are polyclonal antibodies and were extracted from cherry and elderberry respectively. Out of the 17 leaf sample investigated, the antibody CLRV-ch reacted against the CLRV and produced a significant amount of CLRV infection.

The positive controls produced by BIOREBA tested positive. The negative controls provided by BIOREBA also tested positive. This could be due to contamination since negative controls were stored at 4°C for longtime. The detection of CLRV infecting birches by DAS-ELISA has been confirmed by (Nienhaus *et al.*1990), in water (Bandte *et al.*, 2007), root contact (Büttner *et al.*, 2013) and grafting (Cooper *et al.* 1984). Both method of detection (ELISA and RT-PCR) gave positive results for CLRV. The ELISA based techniques often fail, due to low virus titre, during fall and dormant seasons in plant parts like bark and leaves. The presence of various inhibitory compounds in the sap of woody plants may also result in failure of ELISA (Çağlayan, *et al.*, 2006), hence molecular approaches are most sensitive and reliable means for detection of plant viruses, but however time consuming.

For the CLRV strain from elderberry (DAS-ELISA BIOREBA-e), antibodies targeted a low amount of CLRV as reported by (Rebenstorf, 2002). This could be attributed to inability of antibodies to bind with antigen from CLRV strains. The virus particles are blocked from extracts by the coating antibody, and the inhibitory components of extracts are removed by rinsing before addition of detecting antibody and enzyme substrate. The limit of detection of most plant viruses in tests on host tissue extracts is between 1 and 10 ng/ml when using the chromogenic substrate p-nitrophenyl phosphate (NPP) (Van Regenmortel, 1982). The product of hydrolysis of NPP by alkaline phosphate (AP) turns yellow in alkaline solution and strongly absorbs light at 405 nm. Substantially lower limits of detection (>100 nm) can be achieved by using chemiluminescent substrates. Some well plates from (BIOREBA CLRV-e/BIOREBA CLRV-ch) indicated yellow signal but after normalization, they showed signals below the standard threshold.

4.8 Database analysis of the *Betula* sequence data -NCBI

Salojärvi *et al.*, 2019 sequenced 150 birch individuals and assembled *B. pendula* reference genome PRJEB14544 from a fourth-generation in breed line, resulting in a high-quality assembly of 435 Mb that was linked to chromosomes using a dense genetic map. To gain further information on the potential integration of BLRaV into the birch genome *Betula pendula* (PRJEB14544; Salojärvi *et al.*, 2019), *Betula nana* (taxid:216990), *Betula verrucosea* (taxid:3505), a search for additional sequence information of strains from BLRaV as well as Badnaviruses in the whole genome shotgun database at NCBI was conducted as described in chapter 2.12.2.

Upstream and down stream of ORFs provided significant results regarding the existence of partial sequences of badnaviruses (so far unknown sequence information) but not of BLRaV itself as reported already by Rumbou *et al.*, 2020. A special focus on the whole genome shotgun database gave information of badnavirus sequences integrated in *Betula* contigs with partial percentage identity of 70% to BLRaV after blasting using the parameters given in *somehat similar sequence-NCBI* (Fig. 30). There was variability within different strains of BLRaV with regards to different open reading frames (ORF). Some ORF's had high variability than the others. The first step was to summarize the overall variability within different strains of BLRaV as reported by Rumbou *et al.*, 2018.Once the ORF variability is known, the function of a particular ORFs could be predicted by the NCBI.

There are report that vector associated ORFs are in general more variable than other ORFs in *Potato leafroll virus* isolates in solanaceous plants (Guyader *et al.*, 2002). In a first step, the overall variability within different strains of BLRaV needs to be summarized and findout which open reading frame has high variability as well as its function. Results from WGS database after BLASTn of the BLRaV sequence against the organisms; *Betula* (taxid: 3504), *Betula nana* (taxid: 3505), *Betula pendula* (taxid: 3505), *Betula verrucosa* (taxid: 3505) and Betulaceae (taxid: 3514) were filtered according to their length and identity to the particular strain of BLRaV before they were assumed to be relevant. Further investigation was conducted on the potential viral function of regions upstream and downstream of the integration place of the sequence similar to BLRaV by Blastx from the *Betula pendula* genome contig 1188 (Fig. 30).

The whole genome shotgun (WGS) projects are genome assemblies of incomplete genomes or incomplete chromosomes of prokaryotes or eukaryotes that are generally being sequenced by a whole genome shotgun strategy (NCBI, 2021). Although this study provides valuable evidence for a variety of badnavirus sequences in the *Betula* genome database, it leaves several important questions to be answered. Some of the most obvious questions are whether the integration of the sequences found in the genome can be confirmed experimentally in natural *Betula* by other methods. Even though it has been reported that these whole genome sequence could provide bias information; i.e. algorithms just include them in the genome due methodological mistake, it is however interesting to investigate these supposed integrated viruses (NCBI,2021). The upstream and downstream of the open reading frames for whole genome from *Betula pendula* contig 1188 gave significant similarity to Birch leaf roll associated isolate BpubFinlan407501 complete genome with expected value of 0.001 at the target position (ORF+2) and accession number MG68620.1.

A further blastx on this same region gave a functional hit of protein Reverse transcriptase (*Clasdosporium fulvumT-1virus*) at an expected value of 6e-124 (Fig. 31; Tab 16). With these hit, it can be argued that there seems to badnavirus integration in the *Betula pendula* genome. This can be a first step and gives idea to further investigate if these viral hits are endogenous viruses, endogenous viral elements or transposons or retroviral element.

The interpretation of BLRaV integration could be a mistake since the NCBI (whole genome shotgun) has been reported to sometime produce bias result (NCBI, 2021). It is unclear if the process of integration is completely random or whether there are specific sites into which such integration is most likely to occur.

For example, it is known that recombination can occur between badnavirus strains (Ramos-Sobrinho *et al.*, 2020) and might be that, the integration process itself occurs preferentially into recombine genic sequences such as remnants of transposons found widely in plant genomes. The presence of viral sequences, in whole or in part, that are integrated into host genomes, also raises some interesting evolutionary questions. First, depending on the specific insertion site, one would expect that there may be a positive or negative impact on the phenotype of the plant. For any integrated sequences to be maintained actively within the genome, it must be assumed that there is a positive impact on the competitiveness of the host plant.

4.8.1 Genomic analysis of Capillo-like virus sequence alignment

In the genome of *Betula pendula* (contig 934) with accession number FXXK01000935.1, the Capillovirus-like sequence seems to be integrated as reported by Marais *et al.*, 2018. The Blast via NCBI confirmed the presence of Birch capillovirus with accession number of MK402233.1 and MK402234.1 respectively. Upstream and dowstreams of other ORFs proved futile. This gives a first impression of possible integration of Capilloviruses in the *Betula* genome. A hit on the protein level resulted in a poly protein on the ORF1 with a significant E-value of 1e-98. From this limited information it has to be taken into consideration that this Capillovirus-like sequence belongs to the group of endogenous viral elements representing only parts of an ancestral or interrupted viral integrated sequence. An integration long time ago would also explain why only some trees from Wildberg/Nagold did not contain the sequence. This cannot be confirmed yet but reflects a hypothesis. There is therefore a need to find more sequence information for successful completion of the genome and to answer the question if this Capillovirus-like sequence belongs to a true active virus.

5 Badnaviruses state of the art

Badnaviruses have bacilliform shape with double-stranded deoxyribonucleic acid (dsDNA) which are important pathogens in agricultural and horticultural crops in the tropics and temperate regions (Borah *et al.*, 2013). An estimated economic loss of about 10 to 90 % in various crops due to infection from different species of badnaviruses has been reported (Bhat *et al.*, 2016). Out of the eight genera (Caulimovirus, Cavemovirus, Petuvirus, Rosadnavirus, Solendovirus, Soymovirus, Badnavirus and Tungrovirus) within the family *Caulimoviridae*, the genus Badnavirus was given its name because they are bacilliform DNA viruses (Agrios, 2005).

The first badnavirus to be identified was *Commelina yellow mottle virus* (ComYMV) (Migliori and Lastra., 1978; Medberry *et al.*, 1990). After the first discovery, new badnavirus genomes have since evolved and are being characterized based on the available sequences at the nucleotide level with most information being deduced from their nucleotide sequences (Bhat *et al.*, 2016). Even though badnaviruses are generally known to have a single promoter that drives transcription of a terminally redundant pregenomic RNA, there are still less information available with regards to the specific map locations of badnavirus promoters and specific transcription start sites (Tidona and Darai, 2011). For example, both the 5′ and 3′ends of transcripts have been mapped only for ComYMV and *Rice tungro bacilliform virus* (RTBV), whereas only the 5′ end has been mapped for several variants of the *Banana streak virus* (BSV) (Schenk *et al.*, 2001; Medberry *et al.*, 1990; Qu *et al.*, 1991). Data on promoters and transcripts of badnaviruses is an important next step in the characterization of any badnavirus beyond nucleotide sequence level (Tidona and Darai, 2011).

Unlike the retroviruses which have RNA genomes, the integration into host genome is not required for badnaviruses. Instead, they replicate via RNA intermediate by reverse transcription (Geering *et al.*, 2014). They are classified as pararetroviruses because they use the encoded viral reverse transcriptase to replicate their genome (Geering *et al.*, 2014). Although the replication of viral cycle has no integration step, some badnaviruses exist as both episomal (i.e. non-integrated viruses) and as endogenous (i.e. integrated) sequences (endogenous pararetrovirus) in the host plant genome (Staginnus *et al.*, 2009) which can be activated through abiotic stress, giving rise to infective episomal forms (Harper *et al.*, 2002). It is assumed that integration occurs by illegitimate recombination into the host genome, where endogenous pararetroviral sequences may accumulate to high copy numbers (Harper *et al*, 2002; Jakowitsch *et al.*, 1999).

These integrated sequences are assumed to be relics of ancient infection events (Hansen *et al.*, 2005) or representative sequence intermediates between caulimoviruses and retrotransposons (Bousalem *et al.*, 2008). Geering *et al.*, 2014 reported that, there seems to have been a particular trend of integrations based on known endogenous pararetrovirus (EPRV) distribution patterns within and between plant species. Some plant genomes harbor multiple diverse EPRVs, whereas others harbor no known EPRVs; some types of pararetrovirus have been integrated into the genomes of widespread, but distinct, plant species, whereas others have been integrated into a very limited variety of plant genomes (Geering *et al.*, 2014; Staginnus *et al.*, 2009). For instance, three cases infective EPRVs was confirmed by (Chabannes and Iskra-Caruana, 2013) in the respective interspecific hybrids of Tobacco, Petunia and Banana, which suggests that some EPRVs are a reservoir for viral infections. Although the mechanism of integration is poorly understood, nearly all endogenous caulimovirid sequences described so far are fragmented and rearranged when compared with the cognate ancestral viral genome (Geering *et al.*, 2014). It is unlikely to be a coordinated process controlled by the virus, especially as there is no virus-encoded integrase enzyme. Endogenous plant pararetroviruses have received increased attention since their discovery as integrated elements in various plant genomes (Harper *et al*, 2002; Jakowitsch *et al.*, 1999) and their existence poses a new challenge hence the need to provide a reliable diagnostic method for further characterization and management of these pathogens.

5.1 Symptomatology, Host Range and Mode of transmission

Badnaviruses infect both monocotyledonous and dicotyledonous plants with a narrow host range (Bhat *et al.*, 2016). Symptoms caused by badnaviruses normally depend on the developmental stage of host, its cultivars, virus species, as well as biotic and abiotic stress factors. Common badnavirus associated symptoms includes, chlorosis, mottling, necrosis, leaf deformation, and reduced internode length causing stunted growth in some plants (Bhat *et al.*, 2016). Symptoms ranging from mild leaf distortion to death have also been reported to be associated with badnavirus infection (Borah *et al.*, 2013).

For deciduous trees (e.g *Betula* sp.), leaf symptoms of chlorosis and leaf roll has been reported to be associated with *Birch leaf roll associated virus* (BLRaV) (Rumbou *et al.*, 2018). Further investigation on birches from Wildberg/Nagold with symptoms of chlorosis and leaf roll has been associated with BLRaV. Symptomatic birch leaf with abiotic stress factor observed in Gleisdreieck, Berlin has been reported to be infected with BLRaV (Opoku *et al.*, 2018) (Fig. 32).

104

Fig. 32 Symptoms of necrosis and chlorosis due to abiotic stress factor in *Betula* leaves in Gleisdreieck.

Most plants with badnavirus infections often shows asymptomatic and masking features of diseased plants during certain periods (Bhat *et al.*, 2016). Abiotic stress factors such as temperature fluctuation and loss of soil nutrients may cause re-emergence and severity of symptoms. Recent studies using metagenomics approach have also suggested the presence of asymptomatic infections of plants with viruses being more common in nature than previously thought (Pooggin, 2018; Zhang *et al.*, 2018; Kamitani *et al.*, 2016; Barba *et al.*, 2014; Stobbe and Roossinck, 2014; Kreuze *et al.*, 2009). Asymptomatic infections may due to tolerance, in which plants do not suffer from wild type (high titer) virus replication levels, or from viral persistence, in which virus titers are reduced to avoid cytopathic effects and harm to the host. Since a number of definitions of tolerance to viruses exist in genetics, physiology, and ecology, the relation of tolerance to virus titer still remains to be discussed (Råberg, 2014; Little *et al.*, 2010).

Most badnaviruses infecting perennial host plants are transmitted by vegetative propagation with a few ones such as *Commelina yellow mottle virus* (CoYMV), *Kalanchoe top-spotting virus* (KTSV), *Piper yellow mottle virus* (PYMoV), *Cacao swollen shoot virus* (CSSV) and *Taro bacilliform virus* (TaBV) transmitted by seeds (Bhat *et al.,* 2016). Quainoo *et al.*, in 2008 reported the detection of CSSV in every part of Cocoa pod and transmitted through seedlings. A similar successful transmission of the PYMoV by seedling was reported (Deeshma and Bhat, 2014) in black pepper (Piper *nigrum* L.) and TaBV in Taro (*Colocasia esculenta*) by Macanawai *et al.*, 2005.

Fig. 33. Leaf symptoms exhibited on plant species infected by badnavirus: yellow mottle in black pepper (A) (Bhat *et al*., 2016, ICAR-IISR, Kozhikode), leaf roll affecting of birch leaf, (B), necrosis affecting birch leaf (C) (Rumbou *et al*., 2018), mottling affecting leaf of *Ficus carica* (D) (Laney *et al*., 2012), Leaf mosaic and streaks affecting leaf of *Musa sapientum* (Fidan *et al*., 2019).

The secondary mode of transmission of badnaviruses such as, *Rubus yellow net virus* (RYNV) are by aphid vectors (*Amphorophora idaei* and *Amphorophora agathonica)* (Stace-Smith *et al.,*1987*)*, while *Piper yellow mottle virus* (PYMoV) is transmitted by citrus mealybug (*Planococcus citri*) and the black pepper lace bug (*Diconocoris distanti*) in semi persistent manner (Geering, 2014). Mechanical transmission of *Fig badnavirus-1*(FBV-1) to several herbaceous hosts which is known be integrated in the Fig genome. (Laney *et al*., 2012) has also been successful. The top-spotting disease of kalanchoe (*Kalanchoë blossfeldiana*) characterized by numerous yellow spots on the leaves of affected plants are transmitted by mechanical inoculation, grafting, seed and via pollen (Lockhart *et al*., 1988).
Badnaviruses infect number of economically important herbaceous and woody crop species such as *Citrus sp.*, *Fig sp.*, *Grape wine sp.*, *Ribes species, Theobroma cacao* (Borah *et al*., 2013; Bhat *et al*., 2016) as well as *Betula* sp. An example of a genus badnavirus which has been detected in deciduous tree (*Betula pendula* Roth. *and the Betula pubescens* Ehrh.) is the *Birch leaf roll associated virus* (Rumbou *et al*., 2018) which is efficiently transmissible by grafting to non-infected seedlings but has not yet been confirmed to be mechanically transmitted to several herbaceous hosts. Summary of some badnaviruses detected in host plants with segment of genomes and their mode of transmission are listed in (Appendix 11.3. Table A 25).

5.2 Geographical distribution of Badnaviruses

Badnaviruses are regarded as heterogeneous group of viruses, diversed and geographically distributed across tropical and temperate regions of Africa, Asia, Australia, Europe, South and North America (Bhat *et al.*, 2016). The majority of so far known badnavirus species infect tropical and subtropical crops such as Banana, Black pepper, Citrus, Cocoa, Sugarcane, Taro, and Yam (Bhat *et al.*, 2016). Even though badnaviruses are considered to be predominantly distributed in tropical and subtropical origin, a few badnaviruses also infect plants such as Red raspberry, Gooseberry, and Ornamental spiraea in temperate climate zones (Bhat *et al.*, 2016).

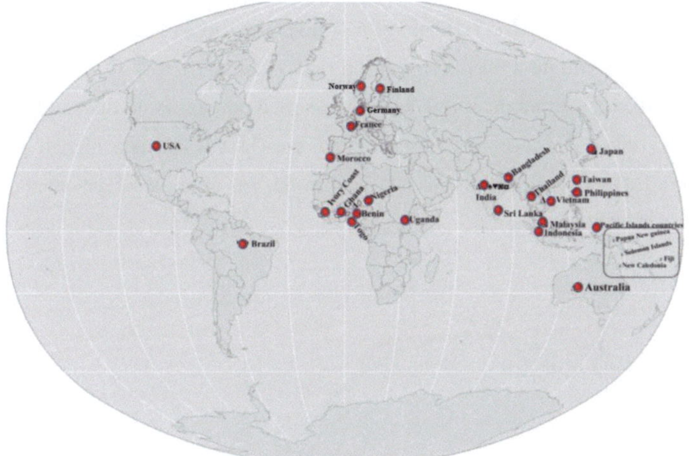

Fig. 34 Geographical distribution of the genus Badnavirus affecting different plant species worldwide. Red spot indicates areas where Badnaviruses been detected (Adapted from Borah *et al.*, 2013).

Recent report has shown the existence of the genus badnavirus with the name *Birch leaf roll associated virus* (BLRaV) infecting forest trees (*Betula* sp.) in some parts of Europe (Rumbou *et al.*, 2018). The symptom in which BLRaV is involved is now described as the Birch leaf-roll disease (BLRD). The BLRD was first recognized as a phytopathological problem in *Betula* spp. in Finland (Jalkanen *et al.*, 2007) but has now been detected in other temperate regions such Germany (Langraf *et al.*, 2016; Opoku *et al.*, 2018), France, Norway, Sweden, Denmark, Switzerland, United Kingdom, and other parts of Europe. The estimation that the genus Badnavirus is associated to tropical and subtropical is currently refuted.

5.3 Genomic organization of badnaviruses

Badnaviruses shares relatively low nucleotide identities when compared to other virus genera (Borah *et al.*, 2013) and normally contain three open reading frames (ORFs) with ORF 1 encoding a protein of unknown function, ORF 2 coding for a virion-associated protein (VAP) and ORF 3 coding for a large polyprotein which is processed into several mature proteins including a movement protein (MP), coat protein (CP), an aspartic protease (AP) reverse transcriptase (RT) and ribonuclease H (RNase H) (Teycheney *et al.*, 2020 ; Geering, 2014) (Fig. 36). Some badnaviruses may have as much as seven open reading frames depending on the type of genus. For instance, *Taro Bacilliform virus* (TaBV) may have four ORF, five (e.g. *Cacao swollen shoot virus*, CSSV), six (e.g. *Citrus yellow mosaic badnavirus*, CMBV) or seven (e.g. *Dracaena sanderiana badnavirus*) (Borah *et al.*, 2013). The badnavirus genome is composed of dsDNA encapsidated into a bacilliform virion with a dimensions of 30 x 130-150 nm and are tubular structures based on a T=3 icosahedron and cut across its threefold axis (Geering, 2014; Hull, 1996). The genome size of the circular dsDNA is between 7.2 and 9.2 Kilo base pair with all the coding capacity on the plus-strand. An example of genome structure of the BLRaV is illustrated in Fig. 35.

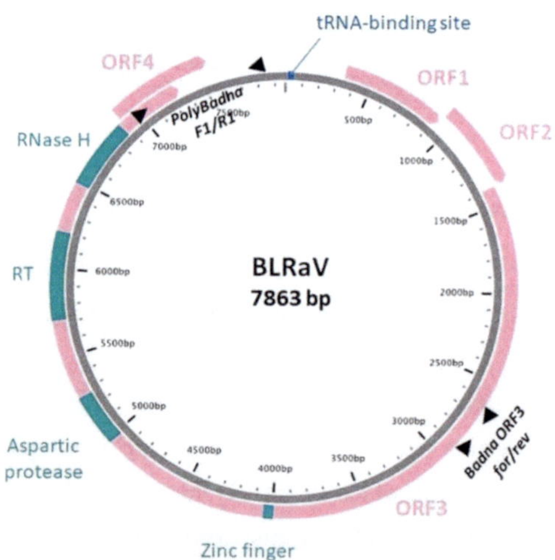

Fig. 35 Genome structure of the *Birch leaf roll associated virus* (BLRaV) showing the positions of the putative tRNAmet-binding site, open reading frames ORF 1; ORF 2; ORF 3 showing movement protein (MP), capsid protein (CP), zinc finger (Zn), aspartic protease (AP), reverse transcriptase (RT) and ribonuclease H (RNase H) motifs (Rumbou *et al.*, 2018).

108

A complete genome sequence of a distinct species BLRaV-badnavirus infecting birches consist of 7500bp with three major open reading frames (ORFs) encoding proteins and a fourth potential ORF overlapping with the end of ORF3 (Rumbou *et al.*, 2018). ORF1, ORF2 and ORF3 showed a low level of amino acid identity to the corresponding proteins encoded by other badnaviruses, reaching a maximum of 44% identity (ORF3) and the *Grapevine vein-clearing virus* appears as the closest badnavirus when considering the polymerase region excluding evidence for presence of endogenous elements.

5.4 Current diagnostic methods for Badnaviruses detection

Bioassays also known as biological indexing is an essential tool used for the evaluation of health status of individual plants, for detection and characterization of a particular virus. In the case of fruit tree viruses, bioassays have increased susceptibility of some fruit tree genotypes to viral infection, resulting in pronounced and easily identifiable symptoms expression. A number of fruit tree genotypes e.g., *Malus platycarpa, Malus micromalus, Malus pumila,* apple and pear virus pathogen have been used for this purpose, (Jelkmann *et al.*, 2001; Damsteegt, *et al.*, 1997). Besides woody indicators, several herbaceous plant species, showing either local or systemic symptoms, have also been used (Legrand *et al.*, 2015; Wood, 1989; Németh *et al.*, 1986) although no universal indicators for all fruit viruses are available. The effectiveness of bioassays varies considerably and depends on several factors. Firstly, there is a need for controlled greenhouse facilities for plant maintenance, due to the vector-mediated transmission of several viruses, thus limiting a large scale and easy-to-use evaluation. Although symptoms on woody indicators could appear few weeks' post-inoculation, they may take from several months to years in the case of some viruses (e.g *Plum pox virus*, PPV) (Maliogka *et al.*, 2020). In addition, some fruit tree viruses with significant levels of intra-species genetic variability may cause symptoms (Woo *et al.*, 2014) as well as responses without any symptoms (e.g., PPV-Rec isolates in GF305 peach seedlings) as reported by Glasa *et al.*, 2004. However, a question arises concerning the health status of used indicators, if not issued from in vitro meristem culture or thermotherapy sanitation, due to the possibility of hidden infections possibly influencing the interpretation of results.

Serological methods such as enzyme-linked immunosorbent assay (ELISA), although limited in sensitivity, enables cheap and fast parallel diagnosis using polyclonal, monoclonal or recombinant antibodies (Terrada *et al.*, 2000; Cambra *et al.*, 1994; Torrance *et al.*, 1981). There has been successful production of antibodies for most herbaceous crops such as *Musa spp.*

However, the production of antibodies against BLRaV affecting *Betula* sp have not yet been successful but successful for *Cherry leaf roll virus* infecting *Betula* sp. The immunogenicity of some fruit tree viruses is scarce, producing antibodies which are not reliable enough for mass-scale diagnosis. Again, the detection of badnaviruses is complicated due to their high serological and genetic heterogeneity (Seal *et al.*, 2014; Muller *et al.*, 2011; Kenyon *et al.*, 2008; Harper *et al.*, 2005). For example, serological methods used for detection of *Banana streak virus* (BSV) was however not successful due to heterogeneity of the BSV isolates until antiserum was produced against a mixture of BSV (Harper *et al.*, 2005; Kenyon *et al.*, 2008; Muller *et al.*, 2011; Seal *et al.*, 2014). Therefore, specific and sensitive detection of most fruit tree viruses and virus-like pathogens is mainly based on molecular diagnosis. The production antibodies are the first step for successful detection.

Polymerase chain reaction as a molecular biological diagnostic tool has successfully been used for detection badnaviruses infecting different herbaceous crops (Boemer *et al.*, 2016) as well as deciduous plants such the *Betula sp.* in Germany (Landgraf *et al.*, 2016; Opoku *et al* 2018). Detection is rapid, sensitive and reliable. *Birch leaf roll associated virus* causing Birch leaf roll disease in *Betula* sp. has been detected by RT-PCR in Finland and some parts of Europe (Rumbou *et al.*, 2018; Opoku *et al.*, 2018; Landgraf *et al.*, 2016). Using specific primers, detection of BLRaV affecting symptomatic birch leaf have been successful at the nucleic acid (RNA level) but not yet successful on the DNA level. Again a rolling circle amplification (RCA) which requires a standard isothermal based assays have been reported for successful detection and amplification of circular double stranded DNA genome of plant pararetroviruses such as *Banana streak virus* (BSV) (Boemer *et al.*, 2016). Since episomal badnavirus genomes are circular (as opposed to endogenous ones), RCA can discriminate between double stranded circular viral genome and endogenous viral sequences and thus overcome false positives, which is a common problem in standard PCR.

High-throughput sequencing (HTS) and bioinformatics are new diagnostic tools used in detection of viruses infecting woody plants. The novel badnavirus (BLRaV) is the first virus to be detected in birches with a complete genome sequence of three viral isolates from Finland and Germany using the RNA- Seq virome analysis (Rumbou *et al.*, 2018). With less information known on plant viruses affecting deciduous woody plants, the HTS has become a reliable tool for further diagnostic and discovery of viral agents in woody hosts which is more reliable than the normal traditional approach (Roossinck *et al.*, 2015, Massart *et al*, 2017).

The use of HTS has accelerated the advancement in hunting for viruses affecting forest and urban trees as well as fruit trees. For example, the RNA sequence analysis in grape wine resulted in production of four different viruses and viroids. Again, up to six viruses and viroids in each of six peach trees (Jo Y *et al.*, 2018) have been detected via HTS. Different approaches are employed in HTS for the detection and identification of plant viruses. One such method is the isolation of total nucleic acid from infected host and massive sequencing (Adams *et al.*, 2009) of total RNA from plant species. De novo sequencing of viruses using deep sequencing is a new technique that has successfully identified known and unknown viruses from long or short reads (Capobianchi *et al.*, 2013).

The extraction of RNA leaves from *Betula* sp. and their diagnostics by metagenomics analysis and deep sequencing in recent years has provided sequences of complex system of virome in birches. One of the results obtained was the BLRaV and there seems to be potential presence of more badnaviruses in the data (Master thesis by Kira Köpke from Division of Phytomedicine-HUB, 2019; Rumbou *et al.*, 2018). Building on these success stories, the application of HTS techniques for routine virus detection has gained momentum. The National center for biotechnology information (NCBI) provides informations on genome of *Betula* sp. which is used as references for identification of BLRaV -badnavirus- associated sequences. BLRaV affecting birches has been successfully detected on the RNA leavel by PCR, hence a need to check at the DNA level using bioinformatics. Information of the DNA level would improve knowledge on the plant genome for further characterization of virus and its pathogenicity

6 Summary

Over the years, birches keep declining at various locations in Europe especially in cities, parks streets as well as forest (Büttner *et al.*, 2022) due to biotic and abiotic stress factors. The decline of birches is characterized by die-back of twigs, branches, and can lead to the loss of tree crown, and finally to death of the tree. Virus-suspected leaf symptoms were observed on declining birches (Landgraf *et al.*, 2017), which were attributed to Birch leaf roll disease (BLRD) (Rumbou *et al.*, 2018). Investigations from this thesis give an impression that, the birches can be infected by more virus species than ever assumed (Opoku *et al.*, 2018). The birch has a complex virome than expected and the decline cannot be attributed not just to a particular virus. Modern techniques, such as high-throughput sequencing (HTS), gave a comprehensive identification of possible viral pathogens present in the birch genome. In 2014, sequences of a previously unknown viruses could be discovered in symptomatic leaf samples of the genus *Betula* (Rumbou *et al.*, 2021). Investigations of the prevalent birch viruses (BLRaV, CLRV, BiCV and ApMV) led to discovery of new viral signatures (e.g. Birch capillovirus, Birch caulimovirus, Birch benyvirus and Birch idaeovirus) in leaf samples from different areas in Europe. This indicates a wide distribution of these viruses as discussed in this thesis. The sequences obtained from complex virome RNA-Seq were about 70 % similar to known plant viruses. With the aid of the high-throughput sequences, RT-PCR-based detection systems for newly discovered viruses have been developed. Furthermore, the genome of these new viruses was to be further characterized and first results on the dissemination of them were presented.

With new viruses evolving from the *Betula*, there is a need to isolate these virus particles (e.g. BLRaV). Even though Serological methods (DAS-ELISA) are available for CLRV detection in birches and have been used for detection of CLRV infecting birches in some selected stands in Berlin, no antibodies for BLRaV detection are known yet. It would be interesting to investigate, if the antibodies can also detect the BLRaV or others from the complex virome since a mechanical transmission of BLRaV was successful. CLRV in birch could be confirmed by RT-PCR as well as serological method, because of the availability of antibodies. What influences the newly found virus signatures such as Birch Benyvirus, Birch caulimovirus and Birch capillovirus have on Birch leaf-roll disease or in ecosystem is still unclear and requires further investigation.

One of the aims of this study was to investigate possible entry pathways from viruses into birch stands. Therefore, a birch seed production site in southern Germany (Wildberg/Nagold) was investigated and a BLRaV infection of mother plus trees was confirmed. These results indicate that, viruses might be introduced from plus tree plantation into the forest and urban ecosystems. Further research on the biology of birch viruses, their host spectra, transmission routes, and their damage potential is needed to derive specific management recommendations. BLRaV could be detected in a leaf sample from BLRD-symptomatic 'mother tree' (plus trees), and a mixed infection could be excluded by HTS. This reinforced and supports the finding of Rumbou *et al.*, 2018 that the symptoms of BLRD infection of birches could be correlated and BLRaV is the potential causal agent of BLRD.

In this context, bioassays were performed to fulfill Koch postulates. BLRaV was successful transmitted on test plants, but a serial passage was not possible. Nevertheless, this partial success is a promising starting point for subsequent studies.

A concept for dealing with virus infected trees is preceded by the identification and detection of the causative pathogens and knowledge of their epidemiology and interactions with the host. A management concept can then be drawn with specific proposals for prophylactic measures to mitigate or eradicate the impact of virus diseases as reported by Fuchs *et al.*, 2020 in grapewine production. The use of certified virus tested propagative material would reduce the inoculum potential especially in areas where vectors are common as reported by Fuchs *et al.*, 2020. The utilization of HTS associated with expertise in bioinformatics is a powerful tool to investigate the *Betula* virome. The application of these knowledge and technologies in birches quarantine and certification programs can improve the efficiency of these programs, contributing to control viral diseases worldwide. Therefore, application of diagnostic methods for the detection of pathogenic viruses in trees should be introduced into the tree management.

7 Zusammenfassung

Seit Jahren gehen Birken an verschiedenen Standorten in Europa, sowohl in Städten an Park-
und Straßenstandplätzen, als auch, in Wäldern, nieder (Buttner *et al.*, 2022). Dieser
Niedergang scheint durch ein Zusammenwirken von biotischen und abiotischen
Stressfaktoren bedingt zu sein. Charakteristisch für den Niedergang ist ein Verkahlen der
Birkenäste und das Absterben dieser, das zu einer Verlichtung der Baumkrone und
schließlich ihrem Verlust führen kann. An niedergehenden Birken wurde virusverdächtige
Blattsymptomen beobachtet (Landgraf *et al.*, 2017), die zum Teilen der
Birkenblattrollkrankheiten (BLRD) zugeschrieben wurden (Rumbou *et al.*, 2018).
Untersuchungen der vorliegenden Arbeit geben einen Eindruck davon, dass Birken von mehr
Virusspezies als bisher angenommen infiziert werden können (Opoku *et al.*, 2018). Die
Birken haben ein komplexeres Virom als erwartet und ihr Niedergang kann nicht nur auf ein
bestimmtes Virus zurückgeführt werden. Moderne Techniken wie die
Hochdurchsatzsequenzierung (HTS) ermöglichten eine zahlreiche Identifizierung möglicher
viraler Krankheitserreger im Microbiom der Birken. Im Jahr 2014, konnten in
symptomatischen Blattproben der Gattung Betula Sequenzen zuvor unbekannter Viren
entdeckt werden. Die weiteren Untersuchungen der weitverbreiten Birkenviren (z. B.
BLRaV, CLRV, BiCV und ApMV) führten zur Entdeckung neuer viraler Signaturen in
Blattproben von Birkenverschiedenen europäischen Standorten. Dies weist auf eine weite
Verbreitung dieser Pflanzenviren hin. Mit Hilfe der Hochdurchsatzsequenzen wurden RT-
PCR-basierende Nachweissysteme für neu entdeckte Viren entwickelt. Außerdem sollte das
Genom dieser neuen Viren weiter charakterisiert und erste Ergebnisse zu ihrer Verbreitung
präsentiert werden. Da sich neue Viren aus *Betula* entwickeln, müssen diese Viren partikel
(z. B. BLRaV) isoliert und charakterisiert werden. Auch wenn serologische Methoden (DAS-
ELISA) zur Verfügung stehen, wäre es interessant zu untersuchen, ob die Antikörper auch
das BLRaV oder andere aus dem komplexen Virom nachweisen können, da eine
mechanische Übertragung von BLRaV erfolgreich war. In Birken konnte CLRV nicht nur
mittels RT-PCR sondern auch durch serologische Verfahren detektiert werden , da
Antikörpern bereits verfügbar sind. Welchen Einfluss die neu gefundenen Viren wie
Benyvirus, Caulimovirus; Capillovirus auf die Birkenblattrollkrankheit oder das Ökosystem
haben, ist noch unklar und bedarf weiterern Untersuchungen. Ein Ziel der vorliegenden
Studie bestand darin mögliche Einbringungswege von Viren in Birkenbestände zu
untersuchen. Dazu wurde eine süddeutsche Birkensamenplantage (Nagold/Wildberg)
erforscht, und eine BLRaV-Infektion von Mutterbäumen wurde bestätigt.

Dieses Ergebnis weist auf, dass Viren möglicherweise von den Samenplantagen in die Wälder und Städte verbracht werden können. Weitere Forschung zur Biologie der Birkenviren, ihren Wirtsspektra, Übertragungswegen und ihrer Schadpotentiale sind nötig um spezifische Managementempfehlungen abzuleiten. Des Weiteren konnte in Blattproben von BLRD-symptomatischen Birken, BLRaV detektiert werden und eine Mischinfektion mittels HTS ausgeschlossen werden. Dies bekräftigt die Feststellung von Rombou et al 2018, dass das Auftreten der BLRD und BLRaV-Infektionen korrelieren und BLRaV das kausale Agens sein könnte. In diesem Zusammenhang wurden Biotest zur Erfüllung der Koch- Postulate durchgeführt. Eine Übertragung des BLRaV war erfolgreich, jedoch war eine serielle Passagierung nicht möglich. Nichtdestotrotz ist dieser Teilerfolg ein erfolgversprechender Ausgangspunkt für fortführende Studien.

Einem Konzept zum Umgang mit virusinfizierten Bäumen gehen die Identifizierung und der Nachweis des ursächlichen Pathogenes und die Kenntnis ihrer Epidemiologie und Wechselwirkungen mit dem Wirt voraus. Anschließend kann ein Managementkonzept mit konkreten Vorschlägen für prophylaktische Maßnahmen zur Minderung der Auswirkungen von Viruserkrankungen erstellt werden, wie von Fuchs *et al.*, 2020 berichtet. Die Verwendung von zertifiziertem virusgetestetem Vermehrungsmaterial würde das Risiko der Virusverbreitung verringern, dies giltinsbesondere in Gebieten in denen Vektoren häufig vorkommen (Fuchs *et al.*, 2020). Die Nutzung von HTS in Verbindung mit bioinformatischen Fachwissen ist ein leistungsfähiges Werkzeug zur Untersuchung des Betula-Viroms. Die Anwendung dieses Wissens und dieser Technologien in Quarantäne- und Zertifizierungsprogrammen für Birken kann die Effizienz dieser Programme verbessern und zur Bekämpfung von Viruserkrankungen weltweit beitragen. Daher sollte die Anwendung diagnostischer Methoden zum Nachweis pathogener Viren in Bäumen in das Management von Forsten und dem Stadtgrün eingeführt werden.

8 References

Adams, I. P., Glover, R. H., Monger, W. A., Mumford, R., Jackeviciene E., Navalinskiene, M., Samuitiene, M., and Boonham, N. (2009). Next-generation sequencing and metagenomic analysis: a universal diagnostic tool in plant virology. Molecular plant pathology 10 (4), 537-545.

Adams, M. J, Candresse, T., Hammond, J., Kreuze, J. F., Martelli, G. P., Namba, S., Pearson, M. N., Ryu, K. H., Saldarelli, P., and Yoshikawa, N. (2012). Betaflexiviridae. In: King AMQ, Adams M. J., Carstens, E. B., Lefkowitz. E. J., (Eds) Virus taxonomy, the Ninth Report of the International Committee on Taxonomy of Viruses. Waltham (USA). Elsevier. 920-941.

Agrios, G.N. (2005) Plant Pathology, Elsevier 5th Edition. 724-725.

Ahamedemujtaba, V., Atheena, P. V., Bhat, A. I., Krishnamurthy, K. S., and Srinivasan, V. (2021). Symptoms of *piper yellow mottle virus* in black pepper as influenced by temperature and relative humidity. Virusdisease 32 (2), 305-313.

Akbas, B., and Degirmenci, K. (2009). Incidence and natural spread of *Apple mosaic virus* on Hazelnut in the west black sea coast of Turkey and its effect on yield. Journal of Plant Pathology 91(3), 767-771.

Almon, E., Horowitz, M., Wang, H. L., Lucas, W. J., Zamski, E., and Wolf, S. (1997). Phloem-specific expression of the *Tobacco mosaic virus* movement protein alters carbon metabolism and partitioning in transgenic potato plants. Plant Physiology, 115, 1599-1607.

Al Rwahnih, M., Daubert, S., Golino, D., Islas, C., and Rowhani, A. (2015). Comparison of next-generation sequencing versus biological indexing for the optimal detection of viral pathogens in grapevine. Phytopathology 105(6), 758-763.

Altschul, S. F, Gish, W., Miller, W., Myers, E. W., and Lipman, D. J. (1990). Basic local alignment search tool. Journal of Molecular Biology, 215 (3), 403-10.

Altschul, S. F, Madden, T. L., Schäffer, A. A, Zhang, J., Zhang, Z., Miller, W., and Lipman D. J. (1997). Gapped BLAST and PSI-BLAST: a new generation of protein database search programs. Nucleic Acids, 25 (17), 3389-402.

Ameyaw, G. A., Dzahini-Obiatey, H. K., and Domfeh, O. (2014). Perspectives on *Cocoa swollen shoot virus disease* (CSSVD) management in Ghana. Crop Protection, 65, 64-70.

Anderson, P. K., Cunningham, A. A., Patel, N. G., Morales, F. J., Epstein, P. R., and Daszak, P. (2004). Emerging infectious diseases of plants: pathogen pollution, climate change and agrotechnology drivers. Trends in Ecology and Evolution 19 (10), 535-544.

Anonymous, (2016). BA Fachbereich Grünflächen.

Anthony Johnson, A. M., Borah, B. K., Sai Gopal, D. V. R., and Dasgupta, I. (2012). Analysis of full-length sequences of two Citrus yellow mosaic badnavirus isolates infecting *Citrus jambhiri* (Rough Lemon) and *Citrus sinensis* L. Osbeck (Sweet Orange) from a nursery in India. Virus Genes, 45(3), 600-605.

Asche, N., Dame, G., Gertz, M., Hein, F., Kreienmeier, U., Leder, B., and Freiherr, V. W. E. (2007). Empfehlungen für die Wiederbewaldung der Orkanflächen in Nordrhein-Westfalen. Landesbetrieb Wald und Holz Nordrhein-Westfalen.

Ashburner, K., and McAllister, H. A. (2013). The Genus *Betula* - A Taxonomic Revision of Birches. Royal Botanic Gardens, Kew.

Balachandran, S., Osmond, C. B. and Daley, R. F. (1994b). Diagnosis of the earliest strain-specific interactions between *Tobacco mosaic virus* and chloroplasts of tobacco leaves in-vivo by means of chlorophyll fluorescence imaging. Plant Physiology. 104:1059-1065.

Bandte, M., and Büttner, C. (2001). A review of an important virus of deciduous trees-*Cherry leaf roll virus*: occurrence, transmission and diagnosis. Pflanzenschutzberichte 59, 1-19.

Bandte, M., Eschevarria-Laza, H. J., Paschek, U., Ulrichs, C., Pestemer, W., Schwarz, D., and Büttner, C. (2007). Transmission of plant viruses by water. In Sociedas Columbiana de Ciencias Horticolas. Proceedings of the 2nd Columbian Congress for Horticulture, Bogota, Colombia (Vol. 1214, p. 3143).

Bandte, M., von Bargen, S., Arndt, N., Grubits, E., Jalkannen, A., and Büttner, C. (2009). Bedeutende Viren an Birken-Fallbeispiele aus Deutschland, Finnland and USA (Virus diseases of birch-occurrence and consequences). Jahrbuch der Baumflege, 2009.

Bandte, M., von Bargen, S., and Büttner, C. (2010). Virusinfektionen an Laubbäumen des öffentlichen Grüns und Forsts Viruses in deciduous trees of public gardens, urban areas and forests. Vom Lebensmittel zum Genussmittel-was essen wir morgen?, 47.

Bandte, M., Schuster, A. K., von Bargen, S., and Büttner, C. (2011). Viren an *Betula pendula* (Roth.) Analyse des Artenspektrums der Ordnung Hemiptera und Nachweis von *Cherry leaf roll virus* (CLRV) in potentiellen Vektoren. In: Jahrbuch der Baumpflege, ed Dujesiefken D,215-221. Haymarket Media: Braunschweig.

Bandte, M., von Bargen, S., Landgraf, M., Rybak, M., and Büttner C. (2022). Neue Erkenntnisse zu Viruserkrankungen an Bäumen und Konsequenzen für die Baumpflege. In: Dujesiefken D. (Ed.), Jahrbuch der Baumpflege 2022, Haymarket Media, Braunschweig, 143-161.

Banerjee, A., Roy, S., and Tarafdar, J. (2012). The large intergenic region of Rice tungro bacilliform virus evolved differentially among geographically distinguished isolates. Virus genes, 44(2), 312-318.

Baranwal, V. K., Arya, M., Singh, J. (2010). First report of two distinct badnaviruses associated with Bougainvillea spectabilis in India. Journal of general plant pathology, 76 (3), 236-239.

Barba, M., Czosnek, H., and Hadidi, A. (2014). Historical perspective, development and applications of next-generation sequencing in plant virology. Viruses, 6(1), 106-136.

Beaver-Kanuya, E., and E., Harper, S.J. (2021). Seasonal fluctuation and host species affect *Tobacco ringspot virus* detection.

Beck, P., Caudullo, G., de Rigo, D. and Tinner, W. (2016). *Betula pendula, Betula pubescens* and other birches in Europe: distribution, habitat, usage and threats. In: San-Miguel-Ayanz, J., de Rigo, D., Caudullo, G., Houston Durrant, T., Mauri, A. (Eds.), European Atlas of Forest Tree Species. Publications Office of the European Union. Luxembourg, e010226+.

Bennetzen., J. L., and Wang, H. (2014). The Contributions of Transposable Elements to the Structure, Function, and Evolution of Plant Genomes. Annual Review of Plant Biology, 65(1), 505-530.

Bhardwaj, P., and Hallan, V. (2019). Molecular evidence of *Apple stem grooving virus* infecting *Ficus palmata*. Trees 33, 1-9.

Bhat, S., Folimonova, S. Y., Cole, A. B., Ballard, K. D., Lei, Z., Watson, B. S., Sumner., L.W., and Nelson, R.S. (2013). Influence of host chloroplast proteins on *Tobacco mosaic virus* accumulation and intercellular movement. Plant Physiology, 161, 134-147.

Bhat, A. I., Hohn, T., and Selvarajan, R., (2016). Badnaviruses: The Current Global Scenario. Viruses, 8(6), 177.

Bhat, A.I., and Rao, G.P. (2020). Transmission of Viruses through Mealybugs. In: Characterization of plant viruses. Springer Protocols Handbooks. Humana, New York, NY.

Biancardi, E., and Tamada, T. (2016). Rhizomania; General Features of *Beet Necrotic Yellow Vein Virus*. (Chapter 3), 55-83.

Boa, E. (2003). An illustrated guide to the state of health of trees recognition and interpretation of symptoms and damage, Diagnostic and Advisory Service; CABI Bioscience, Egham, Surrey, United Kingdom; Food and Agriculture Organization of the United Nations Rome. pp.vi + 49.

Boetzer, M., Henkel, C.V, Jansen, H.J., Butler, D., and Pirovano, W. (2011). Scaffolding pre-assembled contigs using SSPACE, Bioinformatics. 4, 578-579.

Boetzer, M., and Pirovano, W. (2012). Toward almost closed genomes with GapFiller. Genome Biology. 13(6): R56.

Boom, R., Sol C. J., Salimans, M. M., Jansen, C. L., Wertheim-Van dillen, P. M., and van der Noordaa, J. (1990). Rapid and simple method for purification of nucleic acids. Journal of Clinical Microbiology 28 (3), 495-503.

Bos, L. (1977). Symptoms of virus diseases in plants.

Bömer, M., Rathnayake, A.I., Visendi, P. Silva, G., Seal, Susan E. (2018). Complete genome sequence of a new member of the genus Badnavirus, *Dioscorea bacilliform RT virus 3*, reveals the first evidence of recombination in yam badnaviruses. Archives of Virology 163, 533-538.

Borah, B. K., Sharma, S., Kant, R., Johnson, A. M. A., Saigopal, D. V. R., and Dasgupta, I. (2013). Bacilliform DNA-containing plant viruses in the tropics: commonalities within a genetically diverse group. Molecular Plant Pathology, 14 (8), 759-771.

Bousalem, M., Douzery, E. J., and Seal, S. E. (2008). Taxonomy, molecular phylogeny and evolution of plant reverse transcribing viruses (family Caulimoviridae) inferred from full-length genome and reverse transcriptase sequences. Archives of virology, 153(6), 1085-1102.

Breuhahn, M., von Bargen, S., Jalkanen, R., Büttner, C. (2013). Transmission of *Cherry leaf roll virus* (CLRV) variants from German and Finnish birches by grafting. In: Schneider C, Leifert C, Feldmann F (Eds), Endophytes for plant protection: the state of the art, 326-327.

Brookes, A. (2007). Preventing death and serious injury from falling trees and branches. Australian Journal of Outdoor Education, 11, 50-59.

BUND (Der Bund für Umwelt und Naturschutz Deutschland) Baumreport. (2021).

Büttner, C., von Bargen, S., Bandte, M., and Myrta, A. (2011). *Cherry leaf roll virus* chapter 24; Virus and virus-like Diseases of Pome ans stone Fruits: A. Hadidi., M. Barba. T. Candresse and W. Jelkmann. The American Phytopathological Society, 119-125.

Büttner, C., von Bargen, S., Bandte, M., and Mühlbach, H. P. (2013). Forest diseases caused by viruses. Infectious forest diseases, 50-75.

Büttner, C. and Koenig, R. (2013). Viruses. In: Hong, C., Moormann, G., Wohanka, W. and Büttner, C. (eds) Biology, Detection and Management of Plant Pathogens in Irrigation Water. APS Press, St Paul, Minnesota.

Büttner, C., Landgraf, M., Fernández Colino, H.L., von Bargen, S., and Bandte, M. (2022): Editors; Asiegbu, F., and Kovalchuk, A. Forest Microbiology: Tree Diseases and Pests, Volume 3. Virus diseases of forest and urban trees, Chapter 3; 61-98.

Bütow, U., Landgraf, M., Bandte, M., Kube, M., Bergmann, C. C., Behrend, H., Beyerlein, P., and Büttner, C. (2013). Etablierung eines RT-PCR basierenden Nachweises von Cherry leaf roll virus in Birkenpollen (Betula spp.) (Establishment of a RT-PCR based detection method of *Cherry leaf roll virus* in birch pollen (*Betula* spp.). Tagungsbericht 2013, 272.

Çağlayan, K., SERÇE, Ç. U., Gazel, M., and Jelkmann, W. (2006). Detection of four apple viruses by ELISA and RT-PCR assays in Turkey. Turkish Journal of Agriculture and Forestry, 30(4), 241-246.

Cambra, M., Asensio, M., Gorris, M.T., Pérez, E., Camarasa, E., García, J.A., López-Moya, J.J., López-Abella, D., Vela, C. and Sanz, A. (1994). Detection of *Plum pox potyvirus* using monoclonal antibodies to structural and non structural proteins. OEPP/EPPO Bulletin 24: 569-577.

Capobianchi, M. R., Giombini, E., and Rozera, G. (2013). Next-generation sequencing technology in clinical virology. Clinical Microbiology and Infection, 19(1), 15-22.

Card, S.D, Pearson, M.N, Clover, G.R.G. (2007). Plant pathogens transmitted by pollen. Australasian Plant Pathology 36, 455- 461.

Cembali, T., Folwell, R. J., Wandschneider, P., Eastwell, K. C., and Howell, W. E. (2003). Economic implications of a virus prevention program in deciduous tree fruits in the US. Crop Protection, 22(10), 1149-1156.

Chabannes, M., and Iskra-Caruana, M. L. (2013). Endogenous pararetroviruses-a reservoir of virus infection in plants. Current opinion in virology, 3(6), 615-620.

Chang, X., Wang, F., Fang, Q., Chen, F., Yao, H., Gatehouse, A.M.R, and Ye, G. (2021). Virus-induced plant volatiles mediate the olfactory behaviour of its insect vectors. 44 (8), 2700-2715.

Chellappan, P., Vanitharani, R., Ogbe, F., and Fauquet, C. M. (2005). Effect of temperature on geminivirus-induced RNA silencing in plants. Plant physiology, 138(4), 1828-1841.

Chen, J., Shi, Y. H., Lu, Y. W., Adams, M. J., and Chen, J. P. (2006). Narcissus symptomless virus: a new carlavirus of daffodils. Archives of virology, 151(11), 2261-2267.

Chesnais, Q., Caballero Vidal, G., Coquelle, R., Yvon, M., Mauck, K., Brault, V. and Ameline, A., (2020). Post-acquisition effects of viruses on vector behavior are important components of manipulation strategies. Oecologia, 194(3), 429-440.

Chirkov, S., Tsygankova, S., Rastorguev, S., Mitrofanova, I., Chelombit, S., Boulygina, E., Slobodova, N., and Sharko, F. (2021). First report of *Fig cryptic virus* on fig in Russia. *Journal of Plant Pathology*, 103(2), 741-41.

Chung, B.N, Canto, T., Tenllado, F., Choi, K.S., Joa, J.H., Ahn, J.J., Kim, C.H, and Do, K.S. (2016). The Effects of High Temperature on Infection by *Potato virus Y*, *Potato virus A*, and *Potato leafroll virus*. Plant Pathology Journal, 32(4):321-328.

Cieślińska, M. (2020). Differentiation of *Cherry leaf roll virus* isolates from walnut based on molecular analyses and reaction of herbaceous hosts. Physiological and Molecular Plant Pathology, 109, 101456.

Clark, M. F., and Adams, A. N. (1977). Characteristics of the microplate method of enzyme-linked immunosorbent assay for the detection of plant viruses. Journal of general virology, 34(3), 475-483.

Clover, G. R. G., Pearson, M. N., Elliott, D. R., Tang, Z., Smales, T. E., and Alexander, B. J. R. (2003). Characterization of a strain of *Apple stem grooving virus* in *Actinidia chinensis* from China. *Plant Pathology* 52:371-378.

Constable, F. E., Connellan, J., Nicholas, P., and Rodoni, B. C. (2013). The reliability of woody indexing for detection of grapevine virus-associated diseases in three different climatic conditions in Australia. Australian Journal of Grape and Wine Research, 19(1), 74-80.

Cooper, J.I. and Atkinson, M.A. (1975). *Cherry leaf roll virus* causing a disease of *Betula* spp. in the United Kingdom. *Forestry* 48, 193-203.

Cooper, J. I. (1979). Virus diseases of trees and Shrubs. Natural environment research council.

Cooper, J. I., and Massalski P.R. (1984). Viruses and virus-like diseases affecting *Betula* spp. Proceedings of the Royal Society of Edinburgh, Section B: Biological Sciences Volume 85, Issue 1-2 (Birches) 183-195.

Culver, J. N., and Padmanabhan, M. S. (2007). Virus-induced disease: altering host physiology one interaction at a time. Annual review of phytopathology, 45, 221-243.

Czaja, M., Kołton A., and Muras, P. (2020). The complex issue of urban trees-stress factor accumulation and ecological service possibilities. Forests 11(9), 932.

Damsteegt, V. D., Stone, A. L., Mink, G. I., Howell, W. E., Waterworth, H. E., and Levy, L. (1997). The versatility of Prunus tomentosa as a bioindicator of viruses. In XVII International Symposium Virus and Virus-Like Diseases of Temperate Fruit Crops 472.

Dardick, C. (2007). Comparative expression profiling of *Nicotiana benthamiana* leaves systemically infected with three fruit tree viruses. Molecular Plant Microbe Interact 20(8), 1004-1017.

Deeshma, K. P., and Bhat, A. I. (2014). Further evidence of true seed transmission of *Piper yellow mottle virus* in black pepper (*Piper nigrum* L.). Journal of Plant Crops. 42:289-293.

Division of Phytomedicine, HUB (2018). Humboldt University of Berlin

Digiaro, M., and Savino, V. (1992). Role of pollen and seeds in the spread of ilarviruses in almond. Advances in Horticultural Science, 134-136.

Douglas, I. (2008). Environmental change in peri-urban areas and human and ecosystem health. Geography Compass, 2(4), 1095-1137.

Downer, A. J., and Perry, E. J. (2019). Wood decay fungi in landscape trees. UC IPM Pest Notes; UC ANR Publication: Oakland, CA, USA, 74109, 6.

Dubois, H., Verkasalo, E., and Claessens, H., (2020). Potential of Birch (*Betula pendula* Roth and *B. pubescens* Ehrh.) for Forestry and Forest-based Industry sector within the changing climatic and socio-economic context of Western Europe. Forests, 11 (3), 336.

Eastwell, K. C., Du Toit, L. J., and Druffel, K. L. (2009). *Helleborus net necrosis virus*: A New Carlavirus Associated with 'Black Death'of *Helleborus* spp. Plant disease, 93(4), 332-338.

Elena, S. F., Fraile, A., and García-Arenal, F., (2014). Evolution and emergence of plant viruses. Advances in Virus Research. 88, 161-191.

Fidan, H. A. K. A. N., and Koç, G. (2019). Occurrence, ecology and phylogeny of *Banana streak badnavirus* (BSV) and *Cucumber mosaic cucumovirus* (CMV) in *Musa* sp. production areas of the mediterranean coastline of Turkey. Applied Ecology and Environmental Research, 17(3), 5935-5951.

FoVZV *Forstvermehrungsgut-Zulassungsverordnung.* (2002).

Fraile, A., and García-Arenal, F. (2016). Environment and evolution modulate plant virus pathogenesis. *Current Opinion in Virology*, 17, 50-56.

Fuchs, M. (2020). Grapevine viruses: A multitude of diverse species with simple but overall poorly adopted management solutions in the vineyard. Journal of Plant Pathology, 102(3), 643-653.

Fulton, R.W. (1972) *Apple Mosaic Virus* CMI/AAB Descriptions of Plant Viruses, No. 83. Association of Applied Biologists, Wellesbourne.

GALK (*Deutsche Gartenamtsleiterkonferenz*) Straßenbaumliste, (2022). Arbeitskreis Stadtbäume.

Galibina, N. A., Novitskaya, L., Krasavina, M. S., Moshchenskaya, Yu, L. (2015a). Activity of sucrose synthase in trunk tissues of Karelian birch during cambial growth. Russian Journal of Plant Physiology, 62(3): 381-389.

Gambley, C. F., Geering, A. D. W., Steele, V., and Thomas, J. E. (2008). Identification of viral and non-viral reverse transcribing elements in pineapple (Ananas comosus), including members of two new badnavirus species. Archives of Virology, 153(8), 1599-1604.

Gauthier, S., Bernier, P., Kuuluvainen, T., Shvidenko, A.Z., and Schepaschenko, D.G. (2015). Boreal forest health and global change. Science, 349(6250), 819-822.

Geering, A. D. W., Pooggin, M. M., Olszewski, N. E., Lockhart, B. E. L., and Thomas, J. E. (2005). Characterisation of Banana streak Mysore virus and evidence that its DNA is integrated in the B genome of cultivated Musa. Archives of virology, 150(4), 787-796.

Geering, A.D. W., Maumus, F., Copetti, D., Choisne, N., Zwickl, D, Zytnicki, M., McTaggart, A.R., Scalabrin, S., Vezzulli, S., Wing, R.A., Quesneville, H., and Teycheney, P.Y (2014). Endogenous florendoviruses are major components of plant genomes and hallmarks of virus evolution. Nature Communications, 5, 5269.

Geering, A. D. W. (2014). Caulimoviridae (plant pararetroviruses). eLS.

Geering, A. D. W. (2021). Badnaviruses (Caulimoviridae). In Encyclopedia of virology Oxford: Academic Press Elsevier. 158-168.

Gelderblom, H. R. (1996). Structure and classification of viruses. Medical Microbiology. 4th edition.

Gergerich, R. C., and Dolja, V. V., (2006). Introduction to plant viruses, the invisible foe. The plant health instructor, 478.

Gilmer, D., Ratti, C., and Consortium, I. R. (2017). ICTV Virus Taxonomy Profile: Benyviridae. Journal of General Virology, 98, 1571-1572.

Glasa, M., Palkovics, L., Kominek, P., Labonne, G., and Pittnerova, S. Ktidela, O., Candresse, T., Subr, Z. (2004). Geographically and temporally distant natural recombinant isolates of *Plum pox virus* (PPV) are genetically very similar and form a unique PPV subgroup. Journal of General Virology, 85, 2671-2681.

Gotlieb, A.R. and Berbee, J.G. (1973). Line pattern of birch caused by *Apple mosaic virus*. Phytopathology 63, 1470-1477.

Grimová, L., Winkowska, L., Ryšánek, P., Svoboda, P., and Petrzik, K. (2013). Reflects the coat protein variability of *Apple mosaic virus* host preference? Virus Genes 47, 119-125.

Grüntzig, M., Fuchs, E., and Hentsch, T. (1996). Occurrence and serological detection of *Cherry leaf roll nepovirus* (CLRV) and apple mosaic ilarvirus (ApMV) in *Betula* spp. *Zeitschrift für Pflanzenkrankheiten und Pflanzenschutz*, *103*(6), 571-581.

Gutiérrez, S., Michalakis, Y., Van Munster, M., Blanc, S., and Biere, A. (2013). Plant feeding by insect vectors can affect life cycle, population genetics and evolution of plant viruses. Functional Ecology, 27(3), 610-622.

Guyader, S., and Ducray, D. G. (2002). Sequence analysis of Potato leafroll virus isolates genetic stability, major evolutionary events and differential selection pressure between overlapping reading frame products. Journal of General Virology, 83(7), 1799-1807.

Hall, T. A. (1999). BioEdit: a user-friendly biological sequence alignment editor and analysis program for Windows 95/98/NT. In Nucleic acids symposium series (Volume 41, 95-98). (London): Information Retrieval Ltd., c1979-c2000.

Hamacher, J. (1994). *Licht-und elektronenmikroskopische Untersuchungen an Laubbäumen unter verschiedenen Stresseinwirkungen* (Light and electron microscope studies on deciduous trees under different stressconditions). Dr. Kovač.

Hansen, C. N., Harper, G., and Heslop-Harrison, J. S. (2005). Characterisation of pararetrovirus-like sequences in the genome of potato (*Solanum tuberosum*). Cytogenetic and Genome Research, 110 (1-4), 559-565.

Hardcastle, T. and Gotlieb, A.R. (1980). An enzyme-linked immunosorbent assay for the detection of *Apple mosaic virus* in yellow birch. Canadian Journal of Forest Research 10, 278-283.

Harris, K. F., and Maramorosch, K. (Eds.). (2014). Aphids as virus vectors. Elsevier.

Harris, H. M., and Hill, C. (2021). A place for viruses on the tree of life. Frontiers in Microbiology, 11, 604048.

Harper, G., Hull, R., Lockhart, Ben., and Olszewski, N. (2002). Viral sequences integrated into plant genomes. Annual Review of Phytopathology, 40(1), 119-136.

Harper, G., Hart, D., Moult, S., Hull, R., Geering, A., and Thomas, J. (2005). The diversity of *Banana streak virus* isolates in Uganda. Archives of virology, 150(12), 2407-2420.

Hartley, M. J. (2002). Rationale and methods for conserving biodiversity in plantation forests. Forest Ecology Management. 155, 81-95.

Heard, T. A. (1999,). Concepts in insect host-plant selection behavior and their application to host specificity testing. In Host Specificity of Exotic Arthropod Biological Control Agents: The Biological Basis for Improvement in Safety, Proceedings of Session: X International Symposium on Biological Control of Weeds (pp. 1-10).

Heiber, I., Cai, W., and Baier, M. (2014). Linking chloroplast antioxidant defense to carbohydrate availability: the transcript abundance of stromal ascorbate peroxidase is sugar-controlled via ascorbate biosynthesis. Molecular Plant 7(1), 58-70.

Heimonen, K., Valtonen, A., Kontunen, S., Keski-Saari, S., Rousi, M., Oksanen, E., and Roininen, H. (2016). Susceptibility of silver birch (*Betula pendula*) to herbivorous insects is associated with the size and phenology of birch-implications for climate warming. Scandinavian Journal of Forest Research, 32(2), 1-10.

Heinlein, M. (2015). Plant virus replication and movement. Virology, 479, 657-671.

Hemery, G. E., Clark, J. R., Aldinger, E., Claessens, H., Malvolti, M. E., O'connor, E., Y. Raftoyannis, P.S. Savill, R., and Brus, R. (2010). Growing scattered broadleaved tree species in Europe in a changing climate: a review of risks and opportunities. Forestry, 83(1), 65-81.

Henson, J.M. and French, R. (1993). The Polymerase Chain Reaction and Plant Disease Diagnosis. Annual Review of Phytopathology, 31, 81-109.

Heydemann; Hanstein 1984: *Der Forst- und Holzwirt*, p. 536.

Honjo, Mie N., Emura, N., Kawagoe, T., Sugisaka, J., Kamitani, M., Nagano, A.J., and Kudoh, H. (2020). Seasonality of interactions between a plant virus and its host during persistent infection in a natural environment. International Society for Microbial Ecology Journal 14(2), 506-518.

Hull, R. (1996). Molecular biology of rice tungro viruses. Annual review of phytopathology, 34(1), 275-297.

Hull, R. (2001). Matthews' plant virology, 4th Edition., Volume 1. Academic Press, San Diego, California.

Hull, R. (2014). Symptoms and host range. In: Plant virology, 5th Edition. Academic Press, London, 145-198.

Huth, F. (2009). *Untersuchungen zur Verjüngungsökologie der Sand-Birke (Betula pendula Roth)*. PhD-thesis, Faculty of Forest, Geo and Hydro, Dresden University of Technology, Dresden.

Hynynen, J., Niemistö, P., Viherä-Aarnio A., Brunner, A., Hein, S., and Velling, P. (2010). Silviculture of birch (*Betula pendula* Roth and *Betula pubescens* Ehrh.) in northern Europe, Forestry: An International Journal of Forest Research, Volume 83,103-119.

Ingwerson, V. (2021). Bachelor thesis: "*Pollenassoziierte Viren aus Birken in Pollen mit bekanntem Allergengehalt* (PollenPALS DFG BU890/31-1 (655850).

Inouye, N, Maeda, T., and Mitsuhata, K. (1979). *Citrus tatter leaf virus* isolated from Lily. Japanese Journal of Phytopathology, 45(5), 712-720.

International Committee on Taxonomy of Viruses, (ICTV report), (2022).

Jakowitsch, J., Mette, M. F., van Der Winden, J., Matzke, M. A., and Matzke, A. J. (1999). Integrated pararetroviral sequences define a unique class of dispersed repetitive DNA in plants. Proceedings of the National Academy of Sciences of the United States of America, 96(23), 13241-13246.

Jalkanen, R., Büttner, C., and von Bargen, S. (2007). *Cherry leaf roll virus*, CLRV, abundant on *Betula pubescens* in Finland. Silva Fennica, 41, 755-762.

Jelkmann, W. (2001). Detection of Virus and Virus-Like Diseases of Fruit Trees-Laboratory Assays, Bioassays and Indicators. Acta Horticulturae, 1, 473-493.

Johansen, E., Edwards, M. C., and Hampton, R.O. (1994). Seed transmission of Viruses: Current perspectives. Annual Review of Phytopathology, 32(1), 363-386.

Jones, A. T., Koenig, R., Lesemann, D.E., Hamacher, J., Nienhaus, F., and Winter, S. (1990). Serological comparison of isolates of *cherry leafroll virus* from diseased beech, and birch trees in a forest decline area in Germany with other isolates of the virus. Journal of Phytopathology 129: 339-344.

Jones, A. T., McGavin, W. J., Geering, A. D. W., and Lockhart, B. E. L. (2001). A new badnavirus in Ribes species, its detection by PCR, and its close association with gooseberry vein banding disease. Plant Disease, 85(4), 417-422.

Jones, R. A. C. (2009). Plant virus emergence and evolution: origins, new encounter scenarios, factors driving emergence, effects of changing world conditions, and prospects for control. Virus Research. 141(2), 113-130.

Jones, R. A., and Barbetti, M. J. (2012). Influence of climate change on plant disease Infections and epidemics caused by viruses and bacteria. Plant Sciences Reviews, 22, 1-31.

Jones, R. A. C. (2016). Chapter Three - Future Scenarios for Plant Virus Pathogens as Climate Change Progresses, Editor(s): Margaret Kielian, Karl Maramorosch, Thomas C. Mettenleiter, Advances in Virus Research, Academic Press, Volume 95, 87-147.

Jones, R. A.C., and Naidu, R. A. (2019). Global dimensions of plant virus diseases: Current status and future perspectives. Annual review of virology, 6, 387-409.

Junttila, O., and Hänninen, H. (2012). The minimum temperature for budburst in *Betula* depends on the state of dormancy. Tree physiology, 32(3), 337-345.

Kamitani, M., Nagano, A. J., Honjo, M. N., and Kudoh, H. (2016). RNA-Seq reveals virus-virus and virus–plant interactions in nature. FEMS microbiology ecology, 92 (11).

Kanno, Y., Yoshikawa, N., and Takahashi, T. (1993). Some properties of hop latent and apple mosaic viruses isolated from hop plants and their distribution in Japan. Japanese Journal of Phytopathology, 59(6), 651-658.

Kato, S., Shimamoto, Y., and Mikami, T. (1995). The apple mitochondrial atp9 gene: RNA editing and co-transcription with exons a and b of the nad5 gene. *Physiologia Plantarum*, 93(3), 572-575.

Kaur, N., and Hollingsworth (2022). C. Concepts of IPM.

Kenyon, L., Lebas, B. S. M., and Seal, S. E. (2008). Yams (*Dioscorea* spp.) from the South Pacific Islands contain many novel badnaviruses: implications for international movement of yam germplasm. Archives of virology, 153(5), 877-889.

Kibbe, W. A. (2007). OligoCalc: an online oligonucleotide properties calculator. Nucleic acids research, 35(suppl_2), W43-W46.

Kinoti, W. M., Constable, F. E., Nancarrow, N., Plummer, K. M., and Rodoni, B. (2018). The incidence and genetic diversity of Apple mosaic virus (ApMV) and Prune dwarf virus (PDV) in Prunus species in Australia. Viruses, 10(3), 136.

Konijnendijk, C. C., Annerstedt, M., Nielsen, A. B., and Maruthaveeran, S. (2013). Benefits of urban parks. A systematic review. A Report for IFPRA, Copenhagen and Alnarp.

Kouakou, K., Kébé, B. I., Kouassi, N., Ake, S., Cilas, C., and Muller, E. (2012). Geographical distribution of *Cacao swollen shoot virus* Molecular Variability in Côte d'Ivoire. *Plant disease*, 96(10), 1445-1450.

Köpke, K., (2019). Master thesis; Identification of plant virus-associated sequences in high-throughput sequencing data.

Kreuze, J. F., Perez, A., Untiveros, M., Quispe, D., Fuentes, S., Barker, I., and Simon, R. (2009). Complete viral genome sequence and discovery of novel viruses by deep sequencing of small RNAs: a generic method for diagnosis, discovery and sequencing of viruses. *Virology*, 388(1), 1-7.

Landgraf, M., Gehlsen, J., Rumbou, A., Bandte, M., von Bargen, S. Shreiner, M., Jäckel B., and Büttner, C. (2016). *Absterbende Birken im urbanen Grün Berlins – eine Studie zur Virusinfektion. Jahrbuch der Baumpflege* 2016, 276-283.

Landgraf, M., Langer, J., Gröhner, J., Zinnert, L., Bandte, M., von Bargen, S., Schreiner, M., Jäckel B., Büttner, C. (2017). *Viruserkrankungen im urbanen Grün- eine Studie an Birken im Berliner Bezirk Steglitz- Zehlendorf* (Viral diseases in urban areas-a study on birch in Berlin Steglitz-Zehlendorf) Jahrbuch der Baumpflege, 21.

Landgraf, M., Opoku, E. B., Bandte, M., von Bargen, S., Schreiner, M., Jäckel, B., and Büttner, C. (2018). A complex virome identified in declining birch. Vortragsbeitrag; 50. Jahrestagung des DPG-Arbeitskreises Viruskrankheiten der Pflanzen, in Bad Herrenalb.

Langer, J., Schuster, A-K.,von Bargen, S., Bandte, M., and Büttner, C. (2013). Vector transmission of *Cherry leaf roll virus*? Candidate insect species infesting *Betula* spp. In: Schneider C, Leifert C, Feldmann F (Eds), Endophytes for plant protection: the state of the art, 313-314.

Langer, J., Rumbou, A., Fauter, A., von Bargen, S., and Büttner, C. (2016). High genetic variation in a small population of *Cherry leaf roll virus* in *Betula* sp. of montane origin in Corsica. Forest Pathology. Volume 46, Issue 6, 595-599.

Laney, A. G., Hassan, M., and Tzanetakis, I. E. (2012). An integrated badnavirus is prevalent in fig germplasm. Phytopathology, 102(12), 1182-1189.

Lee, P. Y., Costumbrado, J., Hsu, C. Y., and Kim, Y. H. (2012). Agarose gel electrophoresis for the separation of DNA fragments. JoVE (Journal of Visualized Experiments), (62), e3923.

Lefeuvre, P., Martin, D. P., Elena, S. F., Shepherd, D. N., Roumagnac, P., and Varsani, A. (2019). Evolution and ecology of plant viruses. Nature Reviews Microbiology, 17(10), 632-644.

Legrand, P. (2015). Biological assays for plant viruses and other graft-transmissible pathogens diagnoses: a review. EPPO Bulletin, 45(2), 240-251.

Legg, J. P., and Fauquet, C. M. (2004). Cassava mosaic geminiviruses in Africa. *Plant molecular biology*, 56(4), 585-599.

Lehto, K., Tikkanen, M., Hiriart, J.-B., Paakkarinen, V., and Aro, E.-M. (2003). Depletion of the photosystem II core complex in mature tobacco leaves infected by the flavum strain of *tobacco mosaic virus. Molecular Plant-Microbe Interactions* 16(12), 1135-1144.

Lewsey, M., Palukaitis, P., and Carr, J. P. (2009). Plant-virus interactions: defence and counter-defence. Molecular aspects of plant disease resistance, 134-176.

Lheureux, F., Laboureau, N., Muller, E., Lockhart, B. E., and Iskra-Caruana, M. L. (2007). Molecular characterization of banana streak acuminata Vietnam virus isolated from *Musa acuminata siamea* (banana cultivar). Archives of Virology, 152(7), 1409-1416.

Lieutier, F., Yart, A., and Salle, A. (2009). Stimulation of tree defenses by ophiostomatoid fungi can explain attack success of bark beetles on conifers. Annals of Forest Science, 66:801.

Lindenmayer, D. B., and Laurance, W. F. (2017). The ecology, distribution, conservation and management of large old trees. Biological Reviews, 92(3), 1434-1458.

Linnakoski, R., and Forbes, K. M. (2019). Pathogens-The hidden face of forest invasions by wood-boring insect pests. Frontiers in plant science, 10, 90.

Little, T. J., Shuker, D. M., Colegrave, N., Day, T., and Graham, A. L. (2010). The coevolution of virulence: tolerance in perspective. PLoS pathogens, 6(9), e1001006.

Lockhart, B.E.L., and Autrey, L.J.C., (1988). Occurence in sugarcane of a bacilliform virus related serologically to *Banana streak virus*. Plant Disease 72 (3), 230-233.

Lockhart, B. E. L., and Geering, A. D. W. (2000). Partial characterization of two aphid-transmitted viruses associated with yellow leafspot of spiraea. In X International Symposium on Virus Diseases of Ornamental Plants (568), 163-168.

Lorenz, T.C. (2012). Polymerase chain reaction: basic protocol plus troubleshooting and optimization strategies. Journal of Visualized Experiments, (63), e3998.

Louten, J. (2016). Virus structure and classification. Essential human virology, 19.

Lucas, W. J., Olesinski, A., Hull, R. J., Haudenshield, J. S., Deom, C. M., Beachy, R. N., and Wolf, S. (1993). Influence of the *Tobacco mosaic-virus* 30-kda movement protein on carbon metabolism and photosynthate partitioning in transgenic tobacco plants. *Planta* 190(1), 88-96.

Lucas W. J. (2006). Plant viral movement proteins: agents for cell-to-cell trafficking of viral genomes. Virology, 344(1), 169-184.

Macanawai, A. R., Ebenebe, A. A., Hunter, D., Devitt, L. C., Hafner, G. J., and Harding, R. M. (2005). Investigations into the seed and mealybug transmission of *Taro bacilliform virus*. Australasian Plant Pathology, 34(1), 73-76.

Mahlein, A. K., Steiner, U., Hillnhütter, C., Dehne, H. W., and Oerke, E. C. (2012). Hyperspectral imaging for small-scale analysis of symptoms caused by different sugar beet diseases. *Plant methods*, 8(1), 1-13.

Maliogka, V. I., Minafra, A., Saldarelli, P., Ruiz-García, A. B., Glasa, M., Katis, N., and Olmos, A. (2018). Recent advances on detection and characterization of fruit tree viruses using high-throughput sequencing technologies. Viruses, 10(8), 436.

Marais, A., Faure, C., Theil, S., and Candresse, T. (2018). Molecular Characterization of a Novel Species of Capillovirus from Japanese Apricot (*Prunus mume*). *Viruses*, 10(4), 144.

Martelli, G. P. (2017). An overview on grapevine viruses, viroids, and the diseases they cause. Grapevine viruses: *molecular biology, diagnostics and management*, 31-46.

Massart, S., Candresse, T., Gil, J., Lacomme, C., Predajna, L., Ravnikar, M., Reynard, J. S., Rumbou, A., Saldarelli P., Škorić, D., Vainio Eeva J., Valkonen Jari P. T., Vanderschuren, H., Varveri, C., Wetzel, T. (2017). A Framework for the evaluation of biosecurity, commercial, regulatory, and scientific impacts of Plant viruses and viroids identified by NGS technologies. Frontiers in Microbiology, 45.

Mauck, K. E., De Moraes, C. M., and Mescher, M. C. (2014). Biochemical and physiological mechanisms underlying effects of *Cucumber mosaic virus* on host-plant traits that mediate transmission by aphid vectors. *Plant, cell and environment*, 37(6), 1427-1439.

McGrann, G. R., Grimmer, M. K., MUTASA-GÖTTGENS, E. S., and Stevens, M. (2009). Progress towards the understanding and control of sugar beet rhizomania disease. Molecular plant pathology, 10(1), 129-141.

McLeish, M. J., Fraile, A., and García-Arenal, F. (2019). Evolution of plant-virus interactions: host range and virus emergence. Current Opinion in Virology, 34, 50-55.

Medberry, S. L., Lockhart, B. E., Olszewski, N. E. (1990). Properties of *Commelina yellow mottle virus's* complete DNA sequence, genomic discontinuities and transcript suggest that it is pararetrovirus. Nucleic acids research 18(18), 5505-5513.

Menzel, W., Jelkmann, W., and Maiss, E. (2002). Detection of four apple viruses by multiplex RT-PCR assays with coamplification of plant mRNA as internal control. Journal of Virological Methods, 99(1-2), 81-92.

Migliori, A., and Lastra, R. (1978). Study of viruses on *Commelina diffusa* Burm. in Guadeloupe. Institut National de la Recherche Agronomique. In Annales de Phytopathologie (Vol. 10, No. 4, pp. 467-477).

Miljanić, V., Rusjan, D., Škvarč, A., Chatelet, P., and Štajner, N. (2022). Elimination of eight viruses and two viroids from preclonal candidates of six grapevine varieties (Vitis vinifera L.) through in vivo thermotherapy and in vitro meristem tip micrografting. Plants, 11(8), 1064.

Mochizuki, T., Ogata, Y., Hirata, Y., and Ohki, S. T. (2014a). Quantitative transcriptional changes associated with chlorosis severity in mosaic leaves of tobacco plants infected with *Cucumber mosaic virus*. Molecular Plant Pathology. 15(3), 242-254.

Moreno, P., Ambrós, S., Albiach-Marti, M. R. Guerri, J.and Pena, L. (2008). *Citrus tristeza virus*: A pathogen that changed the course of the citrus industry. Molecular plant pathology. 9(2). 251-268.

Moreno, A. B., and López-Moya, J. J. (2020). When Viruses Play Team Sports: Mixed Infections in Plants. Phytopathology, 110(1), 29-48.

Muller, E., Dupuy, V., Blondin, L., Bauffe, F., Daugrois, J. H., Nathalie, L., and Iskra-Caruana, M. L. (2011). High molecular variability of sugarcane bacilliform viruses in Guadeloupe implying the existence of at least three new species. Virus research, 160(1-2), 414-419.

Mülhardt, C. (2010). Molecular biology and genomics. Elsevier.65-94.

National Center for Biotechnology Information (US), and Camacho, C. (2008). BLAST (r) Command Line Applications User Manual (p. 30). National Center for Biotechnology Information (US).

Navarrete-Hernandez, P., and Laffan, K. (2019). A greener urban environment: Designing green infrastructure interventions to promote citizens' subjective wellbeing. Landscape and urban planning, 191, 103618.

Nemeth, M. V. (1992). On the distribution and economic significance of fruit trees in Hungary. Növényvédelem 1992, 28, 26-32.

Németh, M. (1986). Virus diseases of stone fruit trees. Virus, Mycoplasma and Rickettsia Diseases of Fruit Trees, 256-545.

Nienhaus, F. (1985). Infectious diseases in forest trees caused by viruses, mycoplasma-like organisms and primitive bacteria. Experientia 41, 597-603.

Nienhaus, F. and Castello, J.D. (1989). Viruses in forest trees. Annual Review of Phytopathology 27, 165-186.

Nienhaus, F., Büttner, C., and Hamacher, J. (1990). Virus infection of forest trees by mechanical transmission. Journal of Phytopathology 129, 141-150.

Niimi, Y., Han, D.S., Mori, S., and Kobayashi, H. (2003). Detection of *Cucumber mosaic virus*, lily symptomless virus and lily mottle virus in Lilium species by RT-PCR technique. Scientia Horticulturae, 97, 57-63.

Ohsato, S., Miyanishi, M., and Shirako, Y. (2003). The optimal temperature for RNA replication in cells infected by Soil-borne wheat mosaic virus is 17°C. Journal of general virology, 84(4), 995-1000.

Olesinski, A. A., Lucas, W. J., Galun, E., and Wolf, S. (1995). Pleiotropic effects of *Tobacco-mosaic-virus* movement protein on carbon metabolism in transgenic tobacco plants. Planta, 197(1), 118-126.

Olesinski, A. A., Almon, E., Navot, N., Perl, A., Galun, E., Lucas, W. J., and Wolf, S. (1996). Tissue-specific expression of the *Tobacco mosaic virus* movement protein in transgenic potato plants alters plasmodesmal function and carbohydrate partitioning. Plant Physiology, 111(2), 541-550.

Opoku, E. B., Landgraf, M., Pack, K., Bandte, M., von Bargen, S., Schreiner, M., Jackel, B., Büttner, C. (2018). Emerging Plant Viruses in Urban Green: Detection of the Virome in Birch (*Betula* sp.). Journal of Horticulture, 5 (233), 2376-0354.

Otulak, K., Chouda, M., Bujarski, J., and Garbaczewska, G. (2015). The evidence of *Tobacco rattle virus* impact on host plant organelles ultrastructure. Micron 70, 7-20.

Pack, K., Landgraf, M., Opoku, E.B., Bandte, M., von Bargen, S., Artemis Rumbou, A., Schreiner, M., Jäckel, B., and Büttner, C. (2019). Refinement of diagnostic methods for detection of the birch virome using the example of Carla- and Badnaviruses.

Patterson, G.S. (1993). The value of birch in upland forests for wildlife conservation, Forestry Commision Bullettin No. 109. HMSO, London.

Payne, S. (2017). Chapter 10- Introduction to RNA Viruses. Viruses, 97-105.

Petruccelli, R., Bartolini, G., Ganino, T., Zelasco, S., Lombardo, L., Perri, E., Durante, M., and Bernardi, R. (2022). Cold Stress, Freezing Adaptation, Varietal Susceptibility of *Olea europaea* L.: A Review. Plants, 11(10), 1367.

Petrzik, K., Vondrak, J., Peksa, M. O., and Kubesova, O. (2014). Lichens - a new source or yet unknown host of herbaceous plant viruses? European Journal of Plant Pathology 138, 549-559.

Pflanzenschutzamt Berlin; Deutscher Wetterdienst, (2020).

Polák, Z. and Procházková, Z. (1996). *Apple mosaic virus* associated with decline of silver birch. Ochrana Rostlin- UZPI 32, 15-18.

Polák, Z. and Zieglerová, J. (1997). Spontaneous occurrence of *Apple mosaic virus* in some forest and ornamental woody species. In: Cagan, L. and Praslicka, J. (eds) Proceedings of Fourteenth Slovak and Czech Plant Protection Conference (Slovak Republic), 3-4 September 1997. Slovak University of Agriculture, Nitra, Slovakia, pp. 87-88.

Pooggin, M. M., and Ryabova, L. A. (2018). Ribosome Shunting, Polycistronic Translation, and Evasion of Antiviral Defenses in Plant Pararetroviruses and Beyond. Frontiers in microbiology, 9, 644.

Posnette, A. F. (1940). Transmission of swollen shoot disease of cacao. Tropical Agriculture (Trinidad) 17:98.

Possen, J.H.M., Oksanen, E., Rousi, M., Ruhanen, H., Ahonen, V., Tervahauta, A., Heinonen, J., Heiskanen, J., Kärenlampi, S., and Vapaavuori., E. (2011). Adaptability of birch (*Betula pendula* Roth) and aspen (*Populus tremula* L.) genotypes to different soil moisture conditions, *Forest Ecology and Management,* 262, 1387-1399.

Puig, A. S., Wurzel, S., Suarez, S., Marelli, J. P., and Niogret, J. (2021). Mealybug (Hemiptera: Pseudococcidae) Species Associated with *Cacao mild mosaic virus* and Evidence of Virus Acquisition. *Insects*, 12(11), 994.

Qu, R., Bhattacharyya, M., Laco, G. S., De Kochko, A., Rao, B. S., Kaniewska, M. B., Elmer, J.S., Rochestera, D.E., Smith, C.E., and Beachy, R. N. (1991). Characterization of the genome of rice tungro bacilliform virus: comparison with *Commelina yellow mottle virus* and caulimoviruses. Virology, 185(1), 354-364.

Quainoo, A. K., Wetten, A. C., and Allainguillaume, J. (2008). Transmission of *Cocoa swollen shoot virus* by seeds. *Journal of Virological Methods*, 150(1-2), 45-49.

Råberg L. (2014). How to live with the enemy: understanding tolerance to parasites. *PLoS Biology*, 12(11), e1001989.

Ramos-Sobrinho, R., Chingandu, N., A. Gutierrez, O., Marelli, J. P., and K. Brown, J. (2020). A complex of badnavirus species infecting cacao reveals mixed infections, extensive genomic variability, and interspecific recombination. Viruses, 12(4), 443.

Read, A.F. and Taylor, L.H. (2001). The ecology of genetically diverse infections. Science, 292, 1099-1102.

Rebenstorf, K. (2002). Diplomarbeit, Humboldt-Universität Berlin, Landwirtschaftlich-Gärtnerische Fakultät.

Rebenstorf, K. (2005). Untersuchungen zur Epidemiologie des *Cherry leaf roll virus* (CLRV): genetische und serologische Diversität in Abhängigkeit von der Wirtspflanzenart und der geographischen Herkunft. Dissertation Humboldt-Universität zu Berlin, 127 p.

Rebenstorf, K., Candresse, T., Dulucq, M. J., Büttner, C., and Obermeier, C. (2006). Host species-dependent population structure of a pollen-borne plant virus, *Cherry leaf roll virus*. Journal of virology, 80(5), 2453-2462.

Rivas, E. B., Duarte, L. M., Alexandre, M. A. V., Fernandes, F., Harakava, R., and Chagas, C. M. (2005). A new Badnavirus species detected in Bougainvillea in Brazil. Journal of General Plant Pathology, 71(6), 438-440.

Roberts, M., Gilligan, Christopher A., Kleczkowski, A., Hanley, N., Whalley, A. E., Healey, and John R. (2020). The Effect of Forest Management Options on Forest Resilience to Pathogens. Frontiers in Forests and Global Change, 3:7.

Rojas, M. R., and Gilbertson, R. L. (2008). Emerging plant viruses: a diversity of mechanisms and opportunities. In Plant virus evolution (27-51). Springer, Berlin, Heidelberg.

Roloff, A., Gillner, S. (2013). Klimawandel und Stadtbaumarten. Bäume in der Stadt: Besonderheiten, Funktion, Nutzen, Arten, Risiken. Stuttgart: Ulmer, 168-186.

Roossinck, M. J. (2015). Move over, bacteria! Viruses make their mark as mutualistic microbial symbionts. Journal of Virology, 89(13), 6532-6535.

Rott, M., Xiang, Y., Boyes, I., Belton, M., Saeed, H., Kesanakurti, P., Hayes, S., Lawrence, T., Birch, C., Bhagwat, B., and Rast, H. (2017). Application of next generation sequencing for diagnostic testing of tree fruit viruses and viroids. Plant Disease, 101(8), 1489-1499.

Rozen, S. and Skaletsky, H. (2000). Primer3 on the WWW for general users and for biologist programmers. Methods Molecular Biology, 132, 365-386.

Rubio, L., Galipienso, L., and Ferriol, I. (2020). Detection of Plant Viruses and Disease Management: Relevance of Genetic Diversity and Evolution. Frontiers in Plant Science, 11, 1092.

Rumbou, A., von Bargen, S., and Büttner, C. (2009). A model system for plant-virus interaction-infectivity and seed transmission of *Cherry leaf roll virus* (CLRV) in *Arabidopsis thaliana*. European Journal of Plant Pathology, 124. 527-532.

Rumbou, A., von Bargen, S., Demiral, R., Langer, J., Rott, M., Jalkanen, R., and Büttner, C. (2016). High diversity at the inter-/intra-host level of *Cherry leaf roll virus* population associated with the birch leaf-roll disease in Fennoscandia. Scandinavian Journal of Forest Research. 31:6, 546-560.

Rumbou, A., Candresse, T., Marais, A., Theil, S, Lange, J., Jalkanen, R., and Büttner, C., (2018). A novel badnavirus discovered from *Betula* sp. affected by Birch leaf-roll disease. PLoS One; 13 (3), e0193888.

Rumbou, A., Candresse, T., Marais, A., Svanella-Dumas, L., Landgraf, M., Büttner, C. (2020). Unravelling the virome in birch: RNA-Seq reveals a complex of known and novel viruses. PLoS One, 15(6), e0221834.

Rumbou, A., Vainio, E.J., and Büttner, C. (2021). Towards the Forest Virome: High-Throughput Sequencing Drastically Expands Our Understanding on Virosphere in Temperate Forest Ecosystems. Microorganisms 9(8), 1730.

Rush, C. M. (2003). Ecology and epidemiology of benyviruses and plasmodiophorid vectors. Annual review of phytopathology, 41, 567.

Rynkiewicz, Evelyn C., Pedersen, Amy B., and Fenton, A. (2015). An ecosystem approach to understanding and managing within-host parasite community dynamics, Trends in Parasitology, 31(5), 212-221.

Salmond, Jennifer A., Tadaki, M., Vardoulakis, S., Arbuthnott, K., Coutts, A., Demuzere, M., Dirks, Kim N., Heaviside, C., Lim, S., Macintyre, H., McInnes, R.N., and Wheeler, B.W. (2016). Health and climate related ecosystem services provided by street trees in the urban environment. Environmental Health, 15(1),95-111.

Salojärvi, J., Smolander, O. P., Nieminen, K., Rajaraman, S., Safronov, O., Safdari, P., Lamminmäki, A., Immanen, J., Lan, T., Tanskanen, J., Rastas, P., Amiryousefi, A., Jayaprakash, B., Kammonen, J. I., Hagqvist, R., Eswaran, G., Ahonen, V. H., Serra, J. A., Asiegbu, F. O., de Dios Barajas-Lopez, J., and Kangasjärvi, J. (2019). Author correction; Genome sequencing and population genomic analyses provide insights into the adaptive landscape of silver birch. Nature genetics, 51(7), 1187-1189.

Sastry, K. S., and Zitter, T. A. (2014). Management of virus and viroid diseases of crops in the tropics. In Plant virus and viroid diseases in the tropics (149-480). Springer, Dordrecht.

Schenk, P. M., Remans, T., Sági, L., Elliott, A. R., Dietzgen, R. G., Swennen, R., nd Manners, J. M. (2001). Promoters for pregenomic RNA of banana streak badnavirus are active for transgene expression in monocot and dicot plants. Plant Molecular Biology, 47(3), 399-412.

Schoener T. W. (2011). The newest synthesis: Understanding the interplay of evolutionary and ecological dynamics. Science 331(6016), 426-429.

Schmelzer, K. (1972). Nachweis der Verwandtschaft zwischen Herkünften des Kirschenblattroll-Virus *(cherry leaf roll virus)* und dem Ulmenmosaik-Virus (*Elm mosaic virus*) Zentralblatt für Bakteriologie, Parasitenkunde, Infektionskrankheiten und Hygiene (Abteilung 2) 127, 140-144.

Senatsverwaltung für Stadtentwicklung und Umwelt Berlin, Referat Freiraumplanung und Stadtgrün Ausdruck vom 30.06.2016, Bericht Nr. 139, Grünflächeninformationssystem (GRIS) Straßenbäume in Berlin Stand: 31.12.2015 Bestand nach Hauptgattungen in den Berliner Bezirken.

Shahid, M. S., Sattar, M. N., Iqbal, Z., Raza, A., and Al-Sadi, A. M. (2021). Next-generation sequencing and the CRISPR-Cas nexus: A molecular plant virology perspective. Frontiers in Microbiology, 11, 609376.

Sharma, S., and Dasgupta, I. (2012). Development of SYBR Green I based real-time PCR assays for quantitative detection of Rice tungro bacilliform virus and Rice tungro spherical virus. Journal of Virological Methods, 181(1), 86-92.

Shi, B., Lin, L., Wang, S., Guo, Q., Zhou, H., Rong, L., Li, J., Peng, J., Lu, Y., Zheng, H., Yang, Z., Chen, Z., Zhao, J., Jiang, T., Song, B., Chen, J., and Yan, F. (2016). Identification and regulation of host genes related to *Rice stripe virus* symptom production. New Phytologist, 209(3), 1106-1119.

Shiel, P.J., Alrefai, R.H., Domier, L.L., Korban, S.S. and Berger, P. H. (1995). The complete nucleotide sequence of *Apple mosaic virus* RNA-3. Archives of Virology 140, 1247-1256.

Singh, K., Dardick, C., and Kumar Kundu, J. (2019). RNAi-Mediated resistance against viruses in perennial Fruit Plants.Plants 8(10), 359.

Stace-Smith, R., and Jones, A. T., (1987). USDA Agriculture Handbook No.631 (USDA, Washington, p. 175.

Staginnus, C., Iskra-Caruana, M.-L., Lockhart, B. E. L., Hohn, T., and Richert-Poggeler, K. R. (2009). Suggestions for nomenclature of endogenous pararetroviral (EPRV) sequences in plants. Archives Virology. 154:1189-1193.

Stobbe, A. H., and Roossinck, M. J. (2014). Plant virus metagenomics: what we know and why we need to know more. Frontiers in plant science, 5, 150.

Syller J. (2012). Facilitative and antagonistic interactions between plant viruses in mixed infections. Molecular plant pathology, 13(2), 204-216.

Tanner, J. D., Kunta, M., da Graça, J. V., Skaria, M., Nelson, S. D. (2011). Evidence of a Low rate of Seed Transmission of *Citrus tatter leaf virus* in Citrus.

Terrada, E., Kerschbaumer, R. J., Giunta, G., Galeffi, P., Himmler, G., and Cambra, M. (2000). Fully "recombinant enzyme-linked immunosorbent assays" using genetically engineered single-chain antibody fusion proteins for detection of *Citrus tristeza* virus. Phytopathology, 90(12), 1337-1344.

Teshome, D. T., Zharare, G. E., and Naidoo, S. (2020). The Threat of the combined effect of biotic and abiotic stress factors in Forestry under a changing climate. Frontiers in plant science, 11, 601009.

Teskey, R., Wertin, T., Bauweraerts, I., Ameye, M., Mcguire, M. A., and Steppe, K. (2015). Tree response to extreme heat. Plant Cell and Environment, 38, 1699-1712.

Teycheney, P.Y., Geering, A.D.W., Dasgupta, I., Hull, R., Kreuze, J.F., Lockhart, B., Muller, E., Olszewski, N., Pappu, H., Pooggin, M.M., Richert-Pöggeler, K.R., Schoelz, J., Seal, S., Stavolone, L., Umber, M., and ICTV Report Consortium. (2020). ICTV Virus Taxonomy Profile: Caulimoviridae. Journal of General Virology 101:1025-1026.

Thompson, J. D., Higgins, D. G., Gibson, T. J. (1994). CLUSTAL W: improving the sensitivity of progressive multiple sequence alignment through sequence weighting, position-specific gap penalties and weight matrix choice. Nucleic acids research, 22(22), 4673-4680.

Tidona, C., and Darai, G. (2011). The Springer index of viruses: Springer. Berlin, Heidelberg, New York, 2011.

Tidow, S. (2020). Senatsverwaltung für Umwelt, Verkehr und Klimaschutz, Berlin s18-23338.

Tiebel, K, Huth, F.N., and Wagner, S. (2020). Restrictions on natural regeneration of storm-felled spruce sites by silver birch (*Betula pendula* Roth) through limitations in fructification and seed dispersal. European Journal of Forest Research,139(5), 731-745.

Torrance, L., and Jones, R. A. C. (1981). Recent developments in serological methods suited For use in routine testing for plant viruses. Plant Pathology, 30(1), 1-24.

Uhls, A., Petersen, S., Keith, C., Howard, S., Bao, X., and Qiu, W. (2021). Grapevine vein clearing virus is prevalent and genetically variable in grape Aphid (Aphis illinoisensis Shimer) populations. Plant Disease, 105(05), 1531-1538.

Umer, M., Jiwen L., Huafeng, Y., Chuan, X., Kaili, D., Ni Luo, Linghong, K., Xuepei L., Ni H., Guoping W., Xudong F., Ioly, K.L, and Wenxing, Xu. (2019). Genomic, Morphological and Biological Traits of the Viruses Infecting Major Fruit Trees. Viruses, 11(6), 515.

Van Regenmortel, M. H. (1982). Serology and immunochemistry of plant viruses.

Vasquez, D.F., Hernandez, A., Torres, D., Borrero-Echeverry, F., Zuluaga, P. and Rincon, D.F. (2022). Drought as a modulator of plant–virus–vector interactions: Effects on symptom expression, plant immunity and vector behaviour. Plant Pathology, 71,1282-1292.

von Bargen, S., Grubits, E., Jalkanen, R., and Büttner, C. (2009). *Cherry leaf roll virus* – an emerging virus in Finland? Silva Fennica 43,727-738.

von Bargen S., Langer, J., Robel, J., Rumbou, A., Büttner, C. (2012). Complete nucleotide sequence of *Cherry leaf roll virus* (CLRV), a subgroup C nepovirus., 163(2), 0-683.

von Bargen, S., Jalkanen, R., Büttner, C. (2013). Transmission of *Cherry leaf roll virus* (CLRV) variants from German and Finnish birches by grafting. In: Schneider C, Leifert C, Feldmann F (Eds), Endophytes for plant protection: the state of the art, pp. 326-327.

Walker, B.J, Abeel, T., Shea, T., Priest, M., Abouelliel, A., Sakthikumar, S., Cuomo, C.A., Zeng, Q., Wortman, J., Young, S.K., and Earl, A.M. (2014). An integrated tool for comprehensive microbial variant detection and genome assembly improvement. PLoS One. 19; 9(11), e112963.

Walsh, D., and Mohr, I. (2011). Viral subversion of the host protein synthesis machinery. Nature Reviews Microbiology, 9(12), 860-875.

Walthert, L., Ganthaler, A., Mayr, S., Saurer, M., Waldner, P., Walser, M., Zweifel, R., and von Arx, G. (2020). From the comfort zone to crown dieback: Sequence of physiological stress thresholds in mature European beech trees across progressive drought. Science of the total Environment, 141792.

Wells, J.M., H.C. Kirkpatrick, and C.L. Parish (1986). Symptomatology and incidence of *Prunus necrotic ringspot virus* in peach orchards in Georgia. Plant Disease,70(5), 444-447.

Werner, R., Mühlbach, H.P., and Büttner, C. (1997). Detection of *cherry leaf roll nepovirus* (CLRV) in birch, beech and petunia by immuno-capture RT-PCR using a conserved primer pair. Forest Pathology, volume 27(5), 309-318.

Wielkopolan, B., Jakubowska, M., and Obrępalska-Stęplowska A. (2021). Beetles as Plant Pathogen Vectors: Frontiers in Plant Science., 12:748093.

Woo, E. N. Y., and Pearson, M. N. (2014). Biological and molecular variations of *Cherry leaf roll virus* isolates from *Malus domestica*, *Ribes rubrum*, *Rubus idaeus*, *Rumex obtusifolius* and *Vaccinium darrowii*. Plant Pathology, 63(4), 838-845.

Wood, G.A. (1989). A system for propagating and virus-screening imported pome fruit and stone fruit cultivars under cross-hemisphere conditions, New Zealand Journal of Crop and Horticultural Science, 17:2, 169-173.

Yamagishi, N., Sasaki, S., and Yoshikawa, N. (2010). Highly efficient inoculation method of apple viruses to apple seedlings. Julius-Kühn-Archiv, (427), 226.

Yang, I. C., Hafner, G. J., Dale, J. L., and Harding, R. M. (2003). Genomic characterisation of taro bacilliform virus. Archives of Virology, 148(5), 937-949.

Ye, J., McGinnis, S., and Madden, T. L. (2006). BLAST: improvements for better sequence analysis. Nucleic acids research, 34(suppl_2), W6-W9.

Ye, J., Coulouris, G., Zaretskaya, I., Cutcutache, I., Rozen, S., and Madden, T. L. (2012). Primer-BLAST: a tool to design target-specific primers for polymerase chain reaction. *BMC bioinformatics*, *13*(1), 1-11.

Zhang, Y. Z., Shi, M., and Holmes, E. C. (2018). Using Metagenomics to characterize an expanding virosphere. *Cell*, 172(6), 1168-1172.

Zhao, J., Liu, Q., Zhang, H., Jia, Q., Hong, Y., and Liu, Y. (2013). The rubisco small subunit is involved in tobamovirus movement and Tm-22-mediated extreme resistance. Plant Physiology, 161(1), 374-383.

Zhao, J., Zhang, X., Hong, Y., and Liu, Y. (2016). Chloroplast in plant-virus interaction. Frontiers in microbiology, 7, 1565.

144

11 Appendix

11.1 List of Abbreviation

A	Adenine
ACMV	*African cassava mosaic virus*
ApMV	*Apple mosaic virus*
ArMV	*Arabis mosaic virus*
Badnavirus	Bacilliform DNA-Virus
BNYVV	*Beet necrotic yellow vein virus*
BiCV	Birch carlavirus
BIV	Birch idaeovirus
BLAST	Basic Local Alignment Search Tool
BLASTn	Basic Local Alignment Search Tool using a nucleotide query
BLASTx	Basic Local Alignment Search Tool using a translated nucleotide query
BLRaV	*Birch leaf-roll associated virus*
Bp	Basepair
BSc	Bachelor of Science
C	Cytosine
CaMMV	*Cacao mild mosaic virus*
Carlavirus	*Carnation latent virus*
cDNA	Complementary DNA
CLRV	*Cherry leaf roll virus*
Contig	Contiguous
CP	Coat protein
CPRGs	chloroplast and photosynthesis-related genes
CVA	*Cherry virus A*
CuVA	*Currant virus A*
DEPC	Diethyl pyrocarbonate
DdRP	DNA dependant RNA polymerase
dH$_2$O	deionized water
DNA	Deoxyribonucleic Acid
dNTPs	Deoxyribose nucleotide triphosphate
DOP	Division of Phytomedicine
EDTA	Ethylenediamine tetraacetic acid
ELISA	Enzyme-linked immunosorbent assay
EPRV	Endogenous pararetrovirus
F	Forward
GRLHs	Green rice leafhoppers
G	Guanine
HCl	hydrogen chloride
HTS	High-Throughput Sequencing
HU	Universität zu Berlin
HSI	Hyperspectral imaging
ICTV	International Committee on Taxonomy of Viruses
ID	Identity
Kb	Kilobase

MP	Movement protein
mRNA	Messenger RNA
m	Number of cycle
M	Nucleotide
n.a	Not available
NADH	Nicotinamide adenine dinucleotide Hydrogen
Nad5	NADH Dehydrogenase Subunit 5
NCBI	National Center for Biotechnology Information
NGS	Next generation sequencing
Nt	Nucleotide
OD	Optical density
ORF	Open reading frame
PCR	Polymerase chain reaction
POL	Polyemerase
PNRSV	*Prunus necrotic ringspot virus*
PVP	Polyvinylpyrrolidone
PYMoV	*Piper yellow mottle virus*
PYVV	*Potato yellow vein virus*
R	Reverse
RdRP	RNA dependent RNA polymerase
RDV	*Rice dwarf virus*
RNA	Ribonucleic Acid
RNase	Ribonuclease
rpm	Rev per minute
rRNA	ribosomal Ribonucleic Acid
RTase	Reverse Transcriptase
RT-PCR	Reverse transcription polymerase chain reaction
SDS	Sodium dodecyl sulfate
SG	Schwarzer Grund
Sp.	Species
T	Thymine
Taq	Thermus aquaticus
TE	Transposable element
TAE	Tris-Acetate-EDTA
TBE	Tris-Borate-EDTA
TNV	*Tobacco necrosis virus*
TRSV	*Tobacco ringspot virus*
TRSV	*Tomato ringspot virus*
Tris	Tris(hydroxymethyl)aminomethane
tRNA	Transfer RNA
U	units
UTR	Untranslated region
UV	Ultraviolet
V	Volt
v:v	Volume per volume
w:v	Weight per volume
HUB	Humboldt-Universität zu Berlin

Base

A	Adenine
C	Cystosine
G	Guanine
T	Thymine

Mixed Bases (Wobbles)

K	Guanine/Thymine
M	Adenine /Cytosine
R	Adenine /Guanine
S	Cytosine /Guanine
W	Adenine /Thymine
Y	Cytosine/Thymine

11.2   Chemicals

- Bioline

- Biozym

- Carl Roth

- Macherey-Nagel

- Merck

- Thermo Fisher Scientific

- Promega

- Serva

- Sigma Aldrich

- Stratec

All chemicals used in the molecular analysis were produced by above listed companies with their recommended protocols.

11.3 Tables

Table A1. Primers used for RT-PCR detection of viruses with different PCR product size. After BLASTx analysis via NCBI, red nucleotides indicate primers which are not conserved in their regions with accession numbers.

Virus-organism	Primer name and accession numbers	Primer Sequence 5'-3'	Target	Product size	Reference
BLRaV	BadnaSG Forward	ACGAAGAAGGATACCTCACGG	CP	242bp	
	BadnaSG Reverse	CCAGTGTCTAGTACCGCCC			Pack,2016
	BadnaK Forward	GCAGGCACAGAACAATGG		300bp	
	BadnaK Reverse	GGCATTACTAGCCATTCGTA			
BiCV	Carla F	GGCACTTACGTACAGGAGT	CP	197bp	Pack,2016
	(MH536506.1)	GGCACT**C**ACGTACAGGAGT			
	Carla R	GCGGAAAAGGGGCTTAGATA			
	(MH536506.1)	GCGGAA**GAGA**GGCTTAGATA			
CLRV	CLRV-CP350F	CATRGAGAAGCTAAATTTCTCTCT	RNA2-CP	627bp	Demiral, 2014 (MSc.)
	(S63537.1)	**CATAGAGAAGCTAAAATTTCTCTCT**			
	(MK402282.1)	**CATGGAGAAGCTAAAATTTCTCTCT**			
	CLRV-CP977R	ACTCMACCCTATCAAARTATAYCA			
	(MN399681.1)	**ACTCAACCCTATCAAAATATATCA**			
	(MK402282.1)	**ACTCCACCCTATCAAAATATACCA**	RNA2–encoded Polyprotein		von Bargen et al., 2009
	CLRV-CP188 F	TGTCTGTRAATATTATGGCTGGTAC		170bp	
	(MK402282.1)	**TGTCTGTAAATATTATGGCTGGTAC**			
	(KF779179.1)	**TGTCTGTGAATATTATGGCTGGTAC**			
	CLRV-CP350 R	AGAGAGAAATTTAGCTTCTCYATG			
	(S63537.1)	**AGAGAGAAATTTTAGCTTCTCTATG**			
	(MK402282.1)	**AGAGAGAAATTTAGCTTCTCYCTG**			
ApMV	ApMV -CP1739R	GGTGGTAACTCACTCGTTAT	CP	204bp	Langer,2016 (unpublished)
	ApMV -CP1535F	GTAATCCGAAAGGTCCGAAT			
Nad 5(Plant Mitochondria genome)	nad 5 Forward	GATGCTTCTTGGGGCTTCTTGTT	CP	181bp	Menzel et al., 2002
	nad 5 Reverse	CTCCAGTCACCAACATTGGCATAA			

Table A2. RT- PCR components and program for detection of Plant viruses

Table A3. Chemical components for detection of CLRV

Volume (µl)	RT Component
2	cDNA
2.5	10 X PCR buffer without $MgCl_2$ (Phytomedizin)
0.5	dNTPs (10mM each)
0.25	Primer reverse (50 µM)
0.25	Primer forward (50µM)
1	Taq Polymerase (5 u/µl Phytomedizin) (1:20)
2	$MgCl_2$ (25 mM)
16	MilliQ H_2O

Table A4. CLRV – CP350F/CP977R and CLRV-CP 188F/CP350R PCR program. The PCR products were placed on 1.5% agarose gel at 80V for about 45 minutes and observed under UV light using fluorescence dyes (SERVA DNA Stain G).

	Temperature	time	Number of cycles
Primer denaturation	94°C	2 min	1 x
Denaturation	94°C	30 s	
Primer annealing	**53°C**	30 s	30 x
Elongation	72°C	30 s	
Final elongation	72°C	5 min	1 x
	10°C	∞	

Table A5. Chemical components for detection of BadnaSG/Badna K

Volume (µl)	RT Component
2	cDNA
2.5	10 X PCR buffer without $MgCl_2$ (Phytomedizin)
0.5	dNTPs (10mM each)
0.5	**BadnaSG/ R (50 µM)**
0.5	**BadnaSG/ F (50µM)**
1	Taq Polymerase (5u/ µl) (Phytomedizin) (1:20)
2	$MgCl_2$ (25 mM)
16	MilliQ H_2O

Volume (µl)	RT Component
2	cDNA
2.5	10 X PCR buffer without $MgCl_2$(Phytomedizin)
0.5	dNTPs (10mM each)
0.5	**BadnaK/ R (50 µM)**
0.5	**BadnaK/ F (50µM)**
1	Taq Polymerase (5 u/µl) (Phytomedizin) (1:20)
2	$MgCl_2$ (25 mM)
16	MilliQ H_2O

Table A6. BadnaSG PCR program

	Temperature	time	Number of cycles
Primer denaturation	94°C	2 min	1 x
Denaturation	94°C	30 s	
Primer annealing	**55°C**	30 s	30 x
Elongation	72°C	30 s	
Final elongation	72°C	5 min	1 x
	10°C	∞	

Table A7. Badna K- PCR program

	Temperature	Time	Number of cycles
Primer denaturation	94°C	2 min	1 x
Denaturation	94°C	30 s	
Primer annealing	**53°C**	30 s	30 x
Elongation	72°C	30 s	
Final elongation	72°C	5 min	1 x
	10°C	∞	

Table A8. Chemical components for detection of BiCV

Volume (µl)	RT Component
2	cDNA
2.5	10 X PCR buffer without $MgCl_2$(Phytomedizin)
0.5	dNTPs (10mM each)
0.5	**BiCV/R (50 µM)**
0.5	**BiCV/F (50µM)**
1	Taq Polymerase (5 u/ µl) (Phytomedizin) (1:20)
2	$MgCl_2$ (25 mM)
16	MilliQ H_2O

Table A9 BiCV PCR program

	Temperature	Time	Number of cycles
Primer denaturation	94°C	2 min	1 x
Denaturation	94°C	30 s	
Primer annealing	**53°C**	30 s	30 x
Elongation	72°C	30 s	
Final elongation	72°C	5 min	1 x
	10°C	∞	

Table A10 Chemical components for detection of ApMV

Volume (µl)	RT Component
2	cDNA
2.5	10 X PCR buffer without $MgCl_2$ (Phytomedizin)
0.5	dNTPs (10mM each)
0.5	**ApMV -CP1739R (50 µM)**
0.5	**ApMV -CP1535F (50µM)**
1	Taq Polymerase (5 u/µl) (Phytomedizin) (1:20)
2	$MgCl_2$ (25 mM)
16	MilliQ H_2O

Table A11. ApMV PCR program

	Temperature	Time	Number of cycles
Primer denaturation	94°C	2 min	1 x
Denaturation	94°C	30 s	
Primer annealing	**50°C**	30 s	35 x
Elongation	72°C	30 s	
Final elongation	72°C	5 min	1 x
	7°C	∞	

Table A12. Volume and PCR component with primers for detection Benyvirus -like sequences

Volume (µl)	**PCR Component**
2	cDNA
2.5	10 X PCR buffer without $MgCl_2$(Phytomedizin)
0.5	dNTPs (10mM each)
0.25	**Beny 1856F (50 µM)**
0.25	**Beny 1856R (50 µM)**
1	Taq Polymerase (5 u/µl) (Phytomedizin) (1:20)
2	$MgCl_2$ (25 mM)
16.5	MilliQ H_2O
25	**Total volume**

Table A13. PCR program with primers for detection of Benyvirus-like sequences

	Temperature	Time	Number of cycles
Primer denaturation	95°C	2 min	1 x
Denaturation	95°C	20 s	
Primer annealing	**52°C**	20 s	35 x
Elongation	72°C	20 s	
Final elongation	72°C	5 min	1 x
Hold	5°C	∞	

Table A14. PCR component with primers for birch Caulimovirus-like sequence detection

Volume (µl)	**PCR Component**
2	cDNA
2.5	10 X PCR buffer without $MgCl_2$ (Phytomedizin)
0.5	dNTPs (10mM each)
0.25	**Caulimo410A Reverse (50 µl)**
0.25	**Caulimo410A Forward (50 µl)**
1	Taq Polymerase (5 u/µl)(Phytomedizin) (1:20)
2	$MgCl_2$ (25 mM)
16.5	MilliQ H_2O
25	**Total volume**

Table A 15. PCR program with primers for birch Caulimovirus-like sequence detection

	Temperature	Time	Number of cycles
Primer denaturation	94°C	2 min	1 x
Denaturation	94°C	30 s	
Primer annealing	**50°C**	30 s	35 x
Elongation	72°C	30 s	
Final elongation	72°C	5 min	1 x
Hold	5°C	∞	

Table A16. Volume and PCR component with primers for birch Capillovirus-like sequence detection

Volume (µl)	**PCR Component**
2	cDNA
2.5	10 X PCR buffer without MgCl$_2$ (Phytomedizin)
0.5	dNTPs (10mM each)
0.25	**CapilloReverse (50 µM)**
0.25	**CapilloForward (50 µM)**
1	Taq Polymerase (5u/µl) (Phytomedizin) (1:20)
2	MgCl$_2$ (25 mM)
16.5	MilliQ H$_2$O
25	**Total volume**

Table A17. PCR program with primers for birch Capillovirus-like sequence detection

	Temperature	Time	Number of cycles
Primer denaturation	95°C	2 min	1 x
Denaturation	95°C	20 s	
Primer annealing	**53°C**	20 s	35 x
Elongation	72°C	20 s	
Final elongation	72°C	5 min	1 x
Hold	5°C	∞	

Table A 18. Volume and chemical reaction component with primers for birch Deltapartitivirus detection

Volume (µl)	**PCR Component**
2	cDNA
2.5	10 X PCR buffer without MgCl$_2$ (Phytomedizin)
0.5	dNTPs (10mM each)
0.25	**Deltapartiti324Reverse (50 µM)**
0.25	**Deltapartiti324 Forward (50 µM)**
1	Taq Polymerase (5 u/µl) (Phytomedizin) (1:20)
2	MgCl$_2$ (25 mM)
16.5	MilliQ H$_2$O
25	**Total volume**

Table A 19. PCR program for birch Deltapartitivirus detection

	Temperature	Time	Number of cycles
Primer denaturation	94°C	2 min	1 x
Denaturation	94°C	20 s	
Primer annealing	**53°C**	20 s	35 x
Elongation	72°C	20 s	
Final elongation	72°C	5 min	1 x
Hold	5°C	∞	

Table A20

E- numbers, Leaf samples from *Betula* sp. (*B. pendula, pubescens*, and hybrids) collected from roadside, parks and seed plantation have been distinguished based on cataster number from Berlin Grünflächenamt, location based on google maps. Leaf samples were collected in May, June, 2017. Species, concentration of RNA was determined by 2µl Nano drop. Leaf symptoms were distinguished based on symptoms description in chapter 3. Results available are based on RT-PCR according to chapter 2. Positive results are represented by (+) and negative result are represented by (-), N.A (not available), CLRV CP-350F/977R and CLRV CP-188F/ 350R primers were designed based on alignments of CP sequences of different CLRV isolates representing different serotypes (Rebenstorf *et al.* 2006) as described in Gentkow 2010, Böps Schriften, Band 8, 136 pages Cherry leaf roll virus (CLRV): *Charakterisierung ausgewählter Virusisolate unter besonderer Berücksichtigung des viralen Hüllprote*ins. BadnaK and Badna SG are from different strains (Corsica and Schwarzer-Grün), and ApMV.

E-Number	Material	Cataster number	Name of street	Location by Google map	Leaf symptom	Conc. of RNA (ng/µl)	Badna K	Badna SG	Carla K	CLRV- CP 350F/977R	CLRV-CP 188F/350R	ApMV
E54030	Leaf	BI71	Schlangenbader Str.10	52.476189763610 97, 13.308287901701 42	Chlorotic spot	285.7	-	-	-	+	-	-
E54031	Leaf	BI72	Schlangenbader Str.10	52.476189763610 97, 13.308287901701 42	Ringspot	251.4	-	-	-	+	-	-
E54032	Leaf	BI68	Schlangenbader Str.11	52.475017247927 89, 13.307543225877 007	Intercostal chlorosis	192.5	-	-	-	-	-	-
E54033	Leaf	69	Schlangenbader Str.10	52.476189763610 97, 13.308287901701 42	Intercostal chlorosis	273.1	-	-	+	-	-	+
E54034	Leaf	95	Schlangenbader Str.19	52.473933004339 756, 13.307743910464 373	Intercostal chlorosis	157.3	-	-	+	+	-	-
E54035	Leaf	96	Schlangenbader Str.25	52.472885391795 565, 13.308004072721 786	Necrosis	360.3	+	-	+	-	-	-
E54036	Leaf	103	Schlangenbader Str.25	52.472885391795 565, 13.308004072721 786	Necrosis	281.4	+	+	-	-	-	-
E54037	Leaf	104	Schlangenbader Str.25	52.472885391795 565, 13.308004072721 786	Vein banding	259.8	-	-	-	-	-	-
E54038	Leaf	N.A	Altglienicke, 12524 (Rudow)	52.403356376505 144, 13.518203544106 084	Intercostal chlorosis	377.8	-	-	-	-	-	-
E54039	Leaf	N.A	Altglienicke, 12524 (Rudow)	52.403356376505 144, 13.518203544106 084	Leaf deformation	407.9	+	-	-	-	-	+
E54040	Leaf	N.A	Altglienicke, 12524 (Rudow)	52.403356376505 144, 13.518203544106 084	Chlorotic spot	161.7	+	-	-	-	-	+

ID												
E54041	Leaf	N.A	Tempelhofer Feld	52.475468212623554, 13.401840561290513	Oak leaf pattern	217.9	-	-	-	-	-	-
E54042	Leaf	N.A	Tempelhofer Feld	52.475468212623554, 13.401840561290513	Vein banding	442.1	+	-	-	-	-	-
E54043	Leaf	N.A	Tempelhofer Feld	52.475468212623554, 13.401840561290513	Vein banding	214.6	-	-	-	-	-	+
E54044	Leaf	N.A	Tempelhofer Feld	52.475468212623554, 13.401840561290513	Intercostal chlorosis	196.5	+	-	-	-	-	-
E54045	Leaf	N.A	Tempelhofer Feld	52.475468212623554, 13.401840561290513	Intercostal chlorosis	309	+	-	-	-	-	-
E54046	Leaf	I/4273	Verlängerte Rathenaustraße 131a, (Köpenick)	52.463260480106 57, 13.534261498022 145	leaf roll	363.3	+	+	-	-	-	-
E54047	Leaf	II8/197	Verlängerte Rathenaustraße 131a, (Köpenick)	52.463260480106 57, 13.534261498022 145	Necrosis	457.9	+	+	+	-	-	-
E54048	Leaf	II8/196	Verlängerte Rathenaustraße 131a, (Köpenick)	52.463260480106 57, 13.534261498022 145	Intercostal chlorosis	247	-	-	-	-	-	-
E54049	Leaf	III12/235	Verlängerte Rathenaustraße 131a, (Köpenick)	52.463260480106 57, 13.534261498022 145	Chlorosis	604.8	-	-	-	-	-	-
E54050	Leaf	IV1/94	Verlängerte Rathenaustraße 131a, (Köpenick)	52.463260480106 57, 13.534261498022 145	Oak leaf pattern	334.9	+	+	-	-	-	-
E54051	Leaf	M/68	Unter den Birken	52.47121796336362, 13.585906732128306	Intercostal chlorosis	239.3	-	-	-	-	-	-
E54052	Leaf	M/69	Unter den Birken	52.47093510571783, 13.5865111099667	Chlorosis	180.6	+	+	-	-	-	+
E54053	Leaf	M/63	Unter den Birken	52.47065673610103, 13.587056524113544	Necrosis	381.5	-	-	-	-	-	-
E54054	Leaf	M/46	Unter den Birken	52.47065673610103, 13.587056524113544	Leaf roll	370.9	+	+	-	-	-	-
E54055	Leaf	M/42	Unter den Birken	52.47027509747449, 13.587808311180817	Mottling	278.8	-	-	-	-	-	-
E54056	Leaf	N.A	Rabindranath-Tagore-Str. (Grünau)	52.47027509747449, 13.587808311180817	Chlorosis	368.1	-	-	-	-	-	-
E54057	Leaf	134	Rabindranath-Tagore-Str. (Grünau)	52.40793899608361, 13.591038999856139	Vein banding	250.3	-	-	-	-	-	-
E54058	Leaf	202	Rabindranath-Tagore-Str.(Grünau)	52.40793899608361, 13.591038999856139	Chlorosis	435.8	+	-	-	-	-	+
E54059	Leaf	51	Rabindranath-Tagore-Straße (Grünau)	52.40793899608361, 13.591038999856139	Variegation	387.7	+	+	-	+	-	-
E54060	Leaf	64	Rabindranath-Tagore-Straße (Grünau)	52.40793899608361, 13.591038999856139	Mottling	290.2	+	+	-	+	-	-
E54061	Leaf	N.A	Mauerpark (Birkenwäldchen)	52.545679794138955, 13.401190945759282	Mottling	463.4	+	-	-	-	-	-
E54062	Leaf	N.A	Mauerpark (Birkenwäldchen)	52.545689922057676, 13.401158924241523	Oak leaf pattern	403.7	+	+	-	-	-	+
E54063	Leaf	N.A	Mauerpark (Birkenwäldchen)	52.54564147239372, 13.400973993441967	Oak leaf pattern	356.2	-	+	-	-	-	+
E54064	Leaf	N.A	Mauerpark (Birkenwäldchen)	52.54564147239372,	intercostal chlorosis	304.2	+	-	-	-	-	-

158

E54065	Leaf	N.A	Mauerpark (Birkenwäldchen)	13.400973993441967								
E54065	Leaf	N.A	Mauerpark (Birkenwäldchen)	52.54566782897708, 13.40123145479729	Variegation	353.9	+	+	-	-	-	-
E54066	Leaf	N.A	Mauerpark (Birkenwäldchen)	52.54591198477828, 13.401299011488826	Chlorotic spot	304.9	-	+	-	-	-	-
E54067	Leaf	N.A	Mauerpark (Birkenwäldchen)	52.545553306287935, 13.401113363837746	Oak leaf pattern	317	+		-	-	-	-
E54068	Leaf	N.A	Gleisdreick Park	52.494168, 13.377994	Leaf roll	392.5	+	+	-	+	-	-
E54069	Leaf	N.A	Gleisdreick Park	52.494168, 13.377994	leaf roll	231.5	-	+	-	-	-	-
E54070	Leaf	N.A	Gleisdreick Park	52.494168, 13.377994	Chlorosis	230.7	+	-	-	-	-	-
E54071	Leaf	N.A	Gleisdreick Park	52.494168, 13.377994	Intercostal chlorosis	502.8	-	+	-	-	-	-
E54072	Leaf	N.A	Gleisdreick Park	52.493873, 13.374973	Intercostal chlorosis	214	-	-	-	+	-	-
E54073	Leaf	0367	Gleisdreick Park	52.493873, 13.374973	Leaf roll	144.1	-	-	-	+	-	+
E54074	Leaf	N.A	Blankensteinpark	52.522892, 13.461027	Chlorotic spot	486.3	-	-	-	+	-	-
E54075	Leaf	N.A	Blankensteinpark	52.522800, 13.460875	Chlorotic spot	414.2	+	-	-	-	-	+
E54075	Leaf	N.A	Blankensteinpark	52.522800, 13.460875	Chlorotic spot	414.2	+	-	-	-	-	+
E54076	Leaf	N.A	Blankensteinpark	52.522800, 13.460875	Chlorotic spot	608.4	-	+	-	+	-	+
E54077	Leaf	N.A	Blankensteinpark	52.522800, 13.460875	leaf roll	324.5	-	-	-	+	-	+
E54078	Leaf	N.A	Blankensteinpark	52.522647, 13.460703	Leaf deformation	157.4	+	-	-	-	-	+
E54079	Leaf	N.A	Tiergarten (global stone project)	52.512953, 13.373547	Necrosis	314.5	-	+	-	-	-	-
E54080	Leaf	N.A	Tiergarten (global stone project)	52.513159, 13.373263	leaf mosaic	320.4	-	-	-	-	-	-
E54081	Leaf	N.A	Tiergarten (global stone project)	52.513284, 13.373509	leaf mosaic	341.5	+	-	-	-	-	-
E54082	Leaf	N.A	Tiergarten (global stone project)	52.513284, 13.373509	Intercostal chlorosis	410.1	+	+	-	-	-	-
E54083	Leaf	N.A	Tiergarten (global stone project)	52.513430, 13.373915	Chlorotic spot	491.9	-	+	-	-	-	-
E54084	Leaf	30	Wilhelm Gericke Str	52.58403049390777, 13.336332457202085	Variegation	360.7	-	-	-	-	-	-
E54085	Leaf	31	Wilhelm Gericke Str.	52.58392334288665, 13.33655777417188	Mottling	540.2	-	-	-	-	-	+
E54086	Leaf	32	Wilhelm Gericke Str.	52.5833246048268, 13.336648864650282	Vein banding	494.2	-	-	-	-	-	+
E54087	Leaf	33	Wilhelm Gericke Str.	52.58328089909912, 13.336664352640721	Intercostal chlorosis	330.3	-	-	-	-	-	+
E54088	Leaf	34	Wilhelm Gericke Str. (Reineckendorf)	52.58328089909912, 13.336664352640721	Oak leaf pattern	301.5	-	-	+	-	-	-
E54089	Leaf	665	Germanenstraße, Ehrenmal (TreptowerPark)	52.58093157505902, 13.37365593198903	Chlorotic spot	281.8	-	-	+	-	-	-
E54090	Leaf	664	Germanenstraße, Ehrenmal (TreptowerPark)	52.581116802320714, 13.373377444281434	Chlorotic spot	253.4	-	-	+	-	-	-
E54091	Leaf	05123	Germanenstraße, Ehrenmal (TreptowerPark)	52.580757870841985, 13.373863756797746	Mottling	387.6	+	-	+	-	-	-
E54092	Leaf	0407501	Lentzeallee 55/57	52.4684590830746, 13.299326344989336	Necrosis	202.6	+	-	-	-	-	-
E54093	Leaf	0407507	Lentzeallee 55/57	52.4684590830746, 13.299326344989336	Intercostal chlorosis	250	-	+	-	-	-	-
E54094	Leaf	0407510	Lentzeallee 55/57	52.4684590830746, 13.299326344989336	Intercostal chlorosis	306.4	-	+	-	-	-	-

E-number	Material	Tree number	Location	Coordinates	Leaf Symptoms	Conc. of RNA (ng/µl)	Badna K	Badna SG	Carla K	CLRV-CP 350F/977R	CLRV-CP 188F/350R	ApMV
E54095	Leaf	0407519	Lentzeallee 55/57	52.4684590830746, 13.299326344989336	Intercostal chlorosis	282.8	-	+	-	-	-	-
E54096	Leaf	0407521	Lentzeallee 55/57	52.4684590830746, 13.299326344989336	Chlorotic spot	425.6	-	+	+	+	-	-
E54097	Leaf	0407528	Lentzeallee 55/57	52.4684590830746, 13.299326344989336	Variegation	242.7	+	+	-	+	-	-
E54098	Leaf	0407542	Lentzeallee 55/57	52.4684590830746, 13.299326344989336	Variegation	302.5	-	-	-	+	-	-
E54099	Leaf	0407552	Lentzeallee 55/57	52.4684590830746, 13.299326344989336	Leaf deformation	226.5	+	+	-	+	-	-
E54100	Leaf	0407558	Lentzeallee 55/57	52.4684590830746, 13.299326344989336	Intercostal chlorosis	356.1	-	-	-	+	-	-
E54101	Leaf	0407975	Lentzeallee 55/57	52.4684590830746, 13.299326344989336	Leaf roll	332.3	+	+	-	+	-	-

Table A21-Rovaniemi samples -Finland

E- numbers, Leaf samples from *Betula* sp. (*B. pendula, pubescens*, and hybrids) collected from roadside and parks Leaf samples were collected in May, June, 2017. Species, concentration of RNA was determined by 2µl Nano drop. Leaf symptoms were distinguished based on symptoms description in chapter 3. Results are based on RT-PCR according to chapter 2. Positive results are represented by (+) and negative result are represented by (-), n.a (not available), CLRV CP-350F/977R and CLRV CP-188F/ 350R primers were designed base on strains from Corsica (France), BadnaK and Badna SG are from different strains (Corsica and *Schwarz-Grün*), and ApMV.

E-number	Material	Tree number	Location	Leaf Symptoms	Conc. of RNA (ng/µl)	Badna K	Badna SG	Carla K	CLRV-CP 350F/977R	CLRV-CP 188F/350R	ApMV
E56918	Leaf	n.a	Siperia,Hut	Intercostal chlorosis	325.6	+	-	-	-	-	-
E56941	Leaf	51	Ounasvaara Chalets, Rovaniemi	Leaf roll	175.9	+	+	-	-	-	-
E56942	Leaf	52	Ounasvaara Chalets, Rovaniemi	Intercostal chlorosis	178.7	+	+	-	-	-	-
E56943Q	Leaf	53	Ounasvaara Chalets, Rovaniemi	No symptom	292.4	-	-	-	-	-	-
E56943R	Leaf	53	Ounasvaara Chalets, Rovaniemi	No symptom	120.8	-	-	-	-	-	-
E56944	Leaf	54	Ounasvaara Chalets, Rovaniemi	intercostal chlorosis	67.2	+	+	-	-	-	-
E56945	Leaf	55	Antinmukka, Rovaniemi	Ringspot	66.9	-	-	-	-	-	-
E56946	Leaf	56	Pallarintie, Rovaniemie	Vein chlorosis	96.9	+	+	-	-	-	-
E56947	Leaf	3	Pappilantie, Rovaniemi	Leaf roll	249.1	+	-	-	-	-	-
E56948	Leaf	3-I	Pappilantie, Rovaniemi	intercostal chlorosis	125.9	+	+	-	-	-	-
E56949	Leaf	4	Yliopistukatu, Rovaniemi	Vein bending	135.7	+	+	-	-	-	-
E56950	Leaf	5	Yliopistukatu, Rovaniemi	intercostal chlorosis	133.9	+	+	-	+	-	-
E56951	Leaf	6	Yliopistukatu, Rovaniemi	Leaf roll	78.5	+	+	-	-	-	-
E56952	Leaf	7	Katajaranta, Rovaniemi	Vein bending	220.6	+	+	-	-	-	-
E56953	Leaf	8	Ranuantie Rovaniemi	intercostal chlorosis	121.7	+	+	-	-	-	-
E56954	Leaf	8-II	Ranuantie Rovaniemi	Chlorotic spot	92.9	+	+	-	+	-	-
E56955	Leaf	9	Pöykköläntie Rovaniemi	intercostal chlorosis	216.1	+	+	-	-	-	-
E56956	Leaf	9-V	Pöykköläntie Rovaniemi	Necrosis	440	+	+	-	-	-	-
E56957	Leaf	9-VI	Pöykköläntie Rovaniemi	Leaf roll	297.6	+	+	-	-	-	+

E-number	Material	Clone number	Location	Symptom	conc.						
E56958	Leaf	10	Taimelantie Rovaniemi	Intercostal chlorosis	113.1	+	+	-	-	-	-
E56959	Leaf	11	Kaarnikkapolku Rovaniemi	Necrosis	326.2	+	+	-	-	-	-
E56960	Leaf	11-I	Kaarnikkapolku Rovaniemi	Intercostal chlorosis	257.6	+	+	-	-	-	-
E56961	Leaf	11-II	Kaarnikkapolku Rovaniemi	Chlorotic spot	172.9	+	+	-	-	-	-
E56962	Leaf	11-II	Kaarnikkapolku Rovaniemi	No symptom	39.5	-	+	-	-	-	-
E56963	Leaf	20	Eteläranta,Rovaniemi	Intercostal chlorosis	112.6	+	+	-	-	-	-
E56964	Leaf	320	Eteläranta,Rovaniemi	Intercostal chlorosis	165.7	+	+	-	-	-	-

Table A22 Wildberg /Nagold seed plantation samples (Finland)

Overview of *Betula* sp. leaf samples collected from Wildberg/Nagold seed plantation with each tree assigned with an E-number, June 2019. Two specific species (E58128 and E581450 had severe symptoms of leaf roll, vein bending and chlorosis. E- numbers, Leaf samples from *Betula* sp. (*B. pendula, pubescens*, and hybrids) collected from seed plantation. Species, concentration of RNA was determined by 2µl Nano drop. Results available are based on RT-PCR according to chapter 2. Positive results are represented by (+) and negative result are represented by (-), n. a (not available), CLRV CP-350F/977R and CLRV CP-188F/ 350R primers were designed base on strains from Corsica (France), BadnaK and Badna SG are from different strains (Corsica and *Schwarz-Grün*), and ApMV.Primers were designed for Partial viral signatures (DeltapartitiF/R, Caulimo 410AF/R, Benyvirus 1856F/R, capillovirus F/R)

E-number	Clone number	Material	RNA conc.	Badna SG	BadnaK	CLRV CP188F/350R	CLRV CP350F/977R	Carla K	ApMV	Deltapartiti F/R	Caulimo410A F/R	Benyvirus 1856F/R	Capillo virus F/R
E58088	55-04-01	Leaf	949.8	-	-	-	-	-	-	-	+	-	+
E58089	55-04-02	Leaf	627.4	-	-	-	-	-	-	-	+	-	+
E58090	55-04-03	Leaf	723.4	-	-	-	-	-	-	-	+	-	+
E58091	55-04-11i	Leaf	713.9	-	-	-	-	-	-	-	+	-	+
E58092	55-06-01	Leaf	748.8	-	-	-	-	-	-	-	+	-	+
E58093	55-07-01	Leaf	426.1	-	-	-	-	-	-	-	+	-	+
E58094	55-07-02	Leaf	607.1	-	-	-	-	-	-	-	+	-	+
E58095	55-07-03	Leaf	548.5	-	-	-	-	-	-	-	+	-	+
E58096	55-07-04	Leaf	437.1	-	-	-	-	-	-	-	+	-	+
E58097	55-08-01	Leaf	809.8	-	-	-	-	-	-	-	+	-	+
E58098	55-10-01	Leaf	512.4	-	-	-	-	-	-	-	+	-	+
E58099	55-10-02	Leaf	981.4	-	-	-	-	-	-	-	+	-	+
E58100	55-10-03	Leaf	780.0	-	-	-	-	-	-	-	+	-	+
E58101	55-10-04	Leaf	650.01	-	-	-	-	-	-	-	+	-	-
E58102	55-10-05	Leaf	1076.8	-	-	-	-	-	-	-	+	-	+
E58103	55-11-01	Leaf	580.4	-	-	-	-	-	-	-	+	-	+
E58104	55-14-01	Leaf	854.5	-	-	-	-	-	-	-	+	-	+
E58105	55-14-10L	Leaf	1104.4	-	-	-	-	-	-	-	+	-	+
E58106	55-15-01	Leaf	762.4	-	-	-	-	-	-	-	+	-	+
E58107	55-15-02	Leaf	815.2	-	-	-	-	-	-	-	+	-	-
E58108	55-19-01	Leaf	979.6	-	-	-	-	-	-	-	+	-	+

E58109	55-19-02	Leaf	967.1	-	-	-	-	-	-	-	+	-	+
E58110	55-21-01	Leaf	1232.2	-	-	-	-	-	-	-	+	-	+
E58111	55-21-02	Leaf	1004.5	-	-	-	-	-	-	-	+	-	+
E58112	55-21-03	Leaf	1216.9	-	-	-	-	-	-	-	+	-	+
E58113	55-21-04	Leaf	1435.6	-	-	-	-	-	-	-	+	-	+
E58114	55-21-05	Leaf	974.2	-	-	-	-	-	-	-	+	-	+
E58115	55-22-01	Leaf	242.2	-	-	-	-	-	-	-	+	-	-
E58116	55-24-01	Leaf	1432.4	-	-	-	-	-	-	-	+	-	+
E58117	55-24-01A	Leaf	496.8	-	-	-	-	-	-	-	+	-	+
E58118	55-24-01B	Leaf	439.5	-	-	-	-	-	-	-	+	-	+
E58119	55-24-02 plus	Leaf	507.5	-	-	-	-	-	-	-	+	-	-
E58120	55-24-02	Leaf	550.5	-	-	-	-	-	-	-	+	-	-
E58121	55-24-03	Leaf	191.2	-	-	-	-	-	-	-	+	-	+
E58122	55-24-04	Leaf	746.1	-	-	-	-	-	-	-	+	-	+
E58123	55-30-01	Leaf	865.1	-	-	-	-	-	-	-	+	-	+
E58124	55-30-02	Leaf	611.5	-	-	-	-	-	-	-	+	-	+
E58125	55-33-01	Leaf	645.3	-	-	-	-	-	-	-	+	-	-
E58126	55-33-02	Leaf	247.1	-	-	-	-	-	-	-	+	-	+
E58127	55-33-02 plus	Leaf	1027.0	-	-	-	-	-	-	-	+	-	+
E58128	55-35-01	Leaf	1095.1	+	-	-	-	-	-	-	+	-	+
E58129	55-35-02	Leaf	133.2	-	-	-	-	-	-	-	+	-	+
E58130	55-35-03	Leaf	1185.1	-	-	-	-	-	-	-	+	-	+
E58131	55-35-04	Leaf	70.9	-	-	-	-	-	-	-	+	-	+
E58132	55-35-05	Leaf	104.2	-	-	-	-	-	-	-	+	-	+
E58133	55-35-06	Leaf	38.7	-	-	-	-	-	-	-	+	-	-
E58134	55-36-01	Leaf	161.2	-	-	-	-	-	-	-	+	-	-
E58135	55-38-01	Leaf	167.3	-	-	-	-	-	-	-	+	-	-
E58136	55-38-02	Leaf	234.6	-	-	-	-	-	-	-	+	-	-
E58137	55-38-03	Leaf	300.6	-	-	-	-	-	-	-	+	-	-
E58138	55-38-04	Leaf	264.8	-	-	-	-	-	-	-	+	-	-
E58139	55-38-05	Leaf	575.9	-	-	-	-	-	-	-	+	-	+
E58140	55-41-01	Leaf	491.5	-	-	-	-	-	-	-	+	-	+
E58141	55-42-01	Leaf	727.1	-	-	-	-	-	-	-	+	-	-
E58142	55-42-02	Leaf	226.5	-	-	-	-	-	-	-	+	-	+
E58143	55-42-03	Leaf	535.6	-	-	-	-	-	-	-	+	-	-
E58144	55-42-04	Leaf	105.2	-	-	-	-	-	-	-	+	-	+
E58145	55-43-01	Leaf	705.1	+	-	-	-	-	-	-	+	-	+
E58146	55-43-02	Leaf	209.4	-	-	-	-	-	-	-	+	-	+
E58147	55-44-01	Leaf	104.6	-	-	-	-	-	-	-	+	-	+
E58148	55-45-01	Leaf	110.4	-	-	-	-	-	-	-	+	-	+
E58149	55-45-02	Leaf	277.2	-	-	-	-	-	-	-	+	-	+
E58150	55-46-01	Leaf	126.7	-	-	-	-	-	-	-	+	-	+

E58151	55-46-02	Leaf	109.2	-	-	-	-	-	-	-	+	-	+
E58152	55-46-03	Leaf	89.7	-	-	-	-	-	-	-	+	-	+
E58153	55-46-01	Leaf	68.1	-	-	-	-	-	-	-	+	-	+
E58154	55-46-01	Leaf	88.2	-	-	-	-	-	-	-	+	-	+
E58155	55-47-01	Leaf	109.8	-	-	-	-	-	-	-	+	-	+
E58156	55-47-02	Leaf	312.1	-	-	-	-	-	-	-	+	-	+
E58157	55-47-04	Leaf	131.6	-	-	-	-	-	-	-	+	-	-

Table A23 Overview of *Betula* sp. leaf symptoms observed in streets of Berlin from 2015 to 2017. (◻)Observed symptoms; X No observed symptoms. **New extreme symptoms as shown in

Symptoms/Year of rating	2015	2016	2017
Intercostal chlorosis	◻	◻	◻
Chlorosis	◻	◻	◻
Mottling	◻	◻	◻
Leaf roll	◻	◻	◻
Necrosis	◻	◻	◻
Vein bending	◻	◻	◻
Small leaves	◻	◻	◻
Leaf deformation	◻	◻	◻
Chlorotic spot	◻	◻	◻
Mosaic	◻	◻	◻
Line pattern	◻	◻	◻
Ringspot	◻	◻	◻
Variegation	◻	◻	◻
Oak leaf pattern	◻	◻	X
Ringspot**	X	X	◻
Oak leaf pattern**	X	X	◻

Table 24. Standard database and various organisms provided by NCBI used in blasting BLRaV sequences with much focus on Whole genome shotgun (BLASTn) (NCBI, 2019).

Standard database	Organisms
Wgs (whole genome contigs),	**Betulaceae (taxid:3514), *Betula* (taxid:3504), *Betula nana* (taxid:216990), *Betula pendula* (taxid:3505), *Betula verrucosa* (taxid:3505)**
TSA (transcriptome shotgun assembly)	Betulaceae (taxid:3514), *Betula* (taxid:3504), *Betula papyrifera* (taxid:3507), *Betula platyphylla* (taxid:78630): Insects: *Aradus betulae* (taxid:1452765)
SRA (sequence read archive)	20 different organsims: DRX083247 Illumina MiSeq paired end sequencing of SAMD00076717 (*Betula chichibuensis* taxid:312783; study:DRP003742; submission:DRA005642; run:DRR089517); ERX1423598 Illumina HiSeq 2000 sequencing (*Betula nana* taxid:216990; study:ERP001869; submission:ERA600270; run:ERR1352077); ERX1423599 Illumina HiSeq 2000 sequencing (*Betula pubescens* taxid:38787; study:ERP001869; submission:ERA600270; run:ERR1352078); ERX1423600 - ERX1423616 Illumina HiSeq 2000 sequencing (*Betula pubescens* taxid:38787; study:ERP001869; submission:ERA600270; run:ERR1352079) Illumina HiSeq 2000 sequencing (*Betula pubescens* taxid:38787; study:ERP001869; submission:ERA600270; run:ERR1352079)
Est (Expressed sequence tags)	Betulaceae (taxid:3514), *Betula* (taxid:3504), *Betula pendula/Paxillus involutus* mixed EST library (taxid:231415), *Betula platyphylla* (taxid:78630), *Betula pendula* (taxid:3505), *Betula verrucosa* (taxid:3505)
HTGS (High throughput genomic sequences)	Betulaceae (taxid:3514); *Betula* (taxid:3504); *Betula platyphylla* (taxid:78630); *Betula pendula* (taxid:3505); *Betula verrucosa* (taxid:3505); *Biston betularia* (taxid:82595); *Biston betularium* (taxid:82595); *Biston betularius* (taxid:82595); *Conus betulinus* (taxid:89764); *Betula alnobetula* (taxid:38784); *Carpinus betulus* (taxid:12990); *Betula ermanii* (taxid:216992); *Betula lenta* (taxid:216994); *Betula populifolia* (taxid:216989); *Betula nana* (taxid:216990); *Viburnum betulifolium* (taxid:224736); *Betula alba* (taxid:38787); *Betula pubescens* (taxid:38787); *Betula cordifolia* (taxid:1233953); *Betula papyrifera var. cordifolia* (taxid:1233953)
Pat (patent sequences)	Betulaceae (taxid:3514); *Betula* (taxid:3504); *Betula platyphylla* (taxid:78630); *Betula pendula* (taxid:3505); *Betula verrucosa* (taxid:3505); *Betula alnobetula* (taxid:38784); ; *Betula ermanii* (taxid:216992); *Betula lenta* (taxid:216994); *Betula populifolia* (taxid:216989); *Betula nana* (taxid:216990); *Betula alba* (taxid:38787); *Betula pubescens* (taxid:38787); *Betula cordifolia* (taxid:1233953);*Betula papyrifera var. cordifolia* (taxid:1233953); *Betula occidentalis* (taxid:223244); *Betula pendula var. carelica* (taxid:659875); *Betula luminifera* (taxid:312789)
Pdb (Protein data bank)	Betulaceae (taxid:3514); *Betula* (taxid:3504); *Betula platyphylla* (taxid:78630); *Betula pendula* (taxid:3505); *Betula verrucosa* (taxid:3505); *Betula alnobetula* (taxid:38784); *Betula ermanii* (taxid:216992); *Betula lenta* (taxid:216994); *Betula populifolia* (taxid:216989); *Betula nana* (taxid:216990); *Betula alba* (taxid:38787); *Betula pubescens* (taxid:38787); *Betula cordifolia* (taxid:1233953); *Betula papyrifera var. cordifolia* (taxid:1233953)
Refseq_genomic (reference genomic sequences)	Betulaceae (taxid:3514); *Betula* (taxid:3504); *Betula L.* (taxid:3504); *Betula pendula* (taxid:3505); *Betula verrucosa* (taxid:3505); Betula platyphylla (taxid:78630); *Betula nana* (taxid:216990); *Betula lenta* (taxid:216994); *Betula populifolia* (taxid:216989); *Betula alba* (taxid:38787); *Betula pubescens* (taxid:38787); *Betula cordifolia* (taxid:1233953); *Betula papyrifera var. cordifolia* (taxid:1233953); *Betula occidentalis* (taxid:223244)
Gss (genomic survey sequences)	Betulaceae (taxid:3514); Betula (taxid:3504); *Betula platyphylla* (taxid:78630); *Betula pendula* (taxid:3505); *Betula verrucosa* (taxid:3505); *Betula alnobetula* (taxid:38784); *Carpinus betulus* (taxid:12990); *Betula ermanii* (taxid:216992); *Betula lenta* (taxid:216994); *Betula populifolia* (taxid:216989); *Betula nana* (taxid:216990); *Betula alba* (taxid:38787); *Betula pubescens* (taxid:38787); *Betula cordifolia* (taxid:1233953); *Betula papyrifera var. cordifolia* (taxid:1233953)
Dbsts (sequence tagged sites)	Betulaceae (taxid:3514); *Betula* (taxid:3504); *Betula platyphylla* (taxid:78630); *Betula pendula* (taxid:3505); *Betula verrucosa* (taxid:3505); *Betula alnobetula* (taxid:38784); *Betula ermanii* (taxid:216992); *Betula lenta* (taxid:216994); *Betula populifolia* (taxid:216989); *Betula nana* (taxid:216990); *Betula alba* (taxid:38787); *Betula pubescens* (taxid:38787); *Betula cordifolia* (taxid:1233953); *Betula papyrifera var. cordifolia* (taxid:1233953);*Betula occidentalis* (taxid:223244); *Betula pendula var. carelica* (taxid:659875); *Betula luminifera* (taxid:312789)

Table A25 Summary of some badnaviruses detected in host plants with segment of genomes and their mode of transmission.

Name	Abbrev.	Genomic segment	Host	Mode of transmission	References
Banana streak virus	BSV	4	*Musa spp.*	Aphid, mealybug; *D.brevipes, P. citri* and *P. ficus), (S. sacchari*),	Geering *et al.* (2005), Lheureux *et al.* (2007)
Birch leaf roll associated vrius	BLRaV	3	*Betula pendula, B. pubescens*	Grafting	Rumbou *et al.*, 2018
Bougainvillea vein banding-associated badnavirus	BBV	4	*Bougainvillea spectabilis*	unknown	Baranwal *et al.* 2010; Rivas *et al.* (2005)
Cacao swollen shoot virus	CSSV	5	*Theobroma cacao*	Mealybug (*P. citri, P. njalensis F. virgate*) and seed	Kouakou *et al.*,(2012); Quainoo *et al.* (2008)
Citrus yellow mosaic badnavirus	CMBV	6	Citrus	Mealybug (*P. citri*)	Anthony Johnson *et al.* (2012)
Fig badnavirus 1	FBV-1	4	*Ficus carica*	Unknown	Laney *et al.*, 2012
Gooseberry vein banding disease	GVBD	4	*Ribes*	*Phisgrossulariae, Nasonovia ribisnigri,* and *Hyperomyzus* spp. mechanical transmission	Jones *et al.*, 2001
Grapevine vein-clearing virus	GVCV	4	*Vitis spp.*	Grafting, Aphid	Uhls *et al.*, 2021
Pineapple bacilliform comosus virus *Pineapple bacilliform erectifolius virus*	PCV, PBEV	3	*Ananas spp.*	Mealybug(*D. brevipes*)	Gambley *et al.* (2008)
Piper yellow mottle virus	PYMV	Unknown	*Piper nigrum*	Mealybug (*F.virgata, P. citri*) and black pepper lace bug(*D. distant*), Grafting	Bhat *et al* (2013); Geering, (2014)
Rice tungro bacilliform virus	RTBV	4	*Oryza sativa*	Green leafhopper(*Nephotettix virescens, N. cincticeps, resilia dorsalis*)	Banerjee *et al* .(2012); Sharma and Dasgupta (2012)
Spiraea yellow leaf spot virus	SYLSV	-	*Spiraea spp*	Aphid	Lockhart *et al.*,2000
Sugarcane bacilliform virus	SCBV	3	*Saccharum spp.*	Unknown	Lockhart *et al.*, 1988; Muller *et al.* (2011)
Taro bacilliform virus	TaBV	4	*Colocasia esculenta*	Mealy bug	Yang *et al.*,2003
Dioscorea bacilliform RT virus 3	DBRTV3	3	*D.cayenensis,D. bulbifera,D. japonica*	Unknown	Bömer *et al.*, 2018

Table 26 Detection of CLRV affecting birches by DAS-ELISA using CLRV isolates from cherry and elderberry. (-) represent negative detection, (+) represents positive detection and (n.a) represents not applicable.

Samples	E number	Symptoms	BIOREBA CLRV-Ch/DAS ELISA	BIOREBA CLRV-e/DAS ELISA
1	Part of E54094	chlorosis	-	-
2	Part of E54094	Intercostal chlorosis	-	-
3	Part of E54094	Intercostal chlorosis	-	-
4	Part of E54094	Intercostal chlorosis	-	+
5	Part of E54094	Intercostal chlorosis	-	-
6	Part of E54094	Intercostal chlorosis	-	-
7	Part of E54094	Intercostal chlorosis	-	-
8	Part of E54094	Intercostal chlorosis	-	-
9	Part of E54094	Intercostal chlorosis	-	-
10	Part of E54094	Intercostal chlorosis	+	-
(11)510	Part of E54094	Intercostal chlorosis	+	-
(12)519	E54095	Chlorosis	+	+
(13)	407553 BpenMO291B	Ringspot	+	-
(14)	407554 BpenMO291D	Vein banding	+	-
(15)	407555 BpenMO291F	Vein banding	+	-
(16)	407569 BpenMO345B	Intercostal chlorosis	+	-
(17)	407556 BpenMO291H	Intercostal chlorosis	+	-
(18) Positive control	BIOREBA	n.a	+	+
(19) Negative control	BIOREBA	n.a	-	-
(20) Extraction Buffer	n.a	n.a	-	-
(21) Extraction Buffer	n.a	n.a	-	-

11.4 Figures

Figure A1. General overview of birch symptoms; intercostal chlorosis, mottling, vein banding, ringspot, line pattern, chlorotic spot, leaf edge chlorosis, net chlorosis, double ringspot, necrosis, oak leaf pattern and variegation. Necrosis was observed to be the final stage of chlorosis.

Figure A1. General overview of birch symptoms; intercostal chlorosis, mottling, vein banding, ringspot, line pattern, chlorotic spot, leaf edge chlorosis, net chlorosis, double ringspot, necrosis, oak leaf pattern and leaf variegation. Necrosis was observed to be the final stage of chlorosis.

```
                                 10        20        30        40        50        60        70        80        90       100
                                 ....|....|....|....|....|....|....|....|....|....|....|....|....|....|....|....|....|....|....|....|
contig_03284                                                                                                CCCCCCATTGGACGAT
contig_01127                                                                                             ---GCTCGGAAAAAGGC
BetulapubescensGer407526-B5_TA   ACTCCATACAACAAACCCAGAATAAAAAAACCTTCAGATCTTCAATTGACAAAAAAAAAAATGTCTGTAAATCAGAAAATTCGCACAGCTCGGAAAAAGGC

                                 110       120       130       140       150       160       170       180       190       200
                                 ....|....|....|....|....|....|....|....|....|....|....|....|....|....|....|....|....|....|....|....|
contig_03284                     TTGGCATCATTTGACGATTTTGACC CGCCATTGGACCATTTTGATCCACCATTGGACAAAGAAGTGACGAAATTGCCGGATCTGGATGCGAAGGGGAAA
contig_01127                     CGTGGAAGACATGATTTGGGAGAGAATGATAAAGAATCAGAAGAAGACAGAAGGGGGCAGAGAAGTGACGAAATTGCCGGATCTGAATAATTCCGGGGAA
BetulapubescensGer407526-B5_TA   CGTGGAAGACATGATTTGGGAGAGAATGATAAAGAATCAGAAGAAGACAGAAGGGGGCAAAGAAGTGACGAAATTGCCGGATCTGAATACTTCCGGGGAA

                                 210       220       230       240       250       260       270       280       290       300
                                 ....|....|....|....|....|....|....|....|....|....|....|....|....|....|....|....|....|....|....|....|
contig_03284                     GCCCGCGACAAAGCTGATGAGGAGATAATTAGACATTACCTGACGTATGCTTTCGGCAATCTAGCAGTATGGCAAGCTTCAAATGCTGCTGAATTTCCGG
contig_01127                     TCCCGCGACCGAGCTGTTGAGCATATAATTAGACATTACCTCAAACATGCTTTCGGCAATCTAGCAGTATGGCAAGCTTCAAACTCCGCTGAATTTCCGG
BetulapubescensGer407526-B5_TA   TCCCGCGACCGAGCTGTTGAGCATATAATTAGACATTACCTCAAACATGCTTTCGGCAATCTAGCAGTATGGCAAGCTTCAAACTCCGCTGAATTTCCGG

                                 310       320       330       340       350       360       370       380       390       400
                                 ....|....|....|....|....|....|....|....|....|....|....|....|....|....|....|....|....|....|....|....|
contig_03284                     ATGTAAATATCAATTTGCCAAATGGAATTTACGGTTCAAAAGTTTGGGACGATCAGGTTGAGGTGAGTCTTAATATGAACATCTTCGTAAGGTGCCTCAG
contig_01127                     ATGTGAAGATCTATTTGCCAACTGTAACTTACAGAATCAAAAACAGGGAAAATCGGGCTAATCTCAGTCTTAATATGAAGAACTTCGTAAGCTGCCTCAG
BetulapubescensGer407526-B5_TA   ATGTGAAGATCTATTTGCCAACTGTAACTTACAGAATCAAAAACAGGGAAAATCGGGCTAAGCTCAGTCTTAATATGAAGAACTTCGTAAGCTGCCTCAG

                                 410       420       430       440       450       460       470       480       490       500
                                 ....|....|....|....|....|....|....|....|....|....|....|....|....|....|....|....|....|....|....|....|
contig_03284                     GGAAGTTAGTAGAACGAGCAGTAATATTGTTGTCAATGGGGCCACGATCGTGCAACTAGCAGAGCCTTTCGCTGATTTAGCAGTACGGTATTTGAGAT-
contig_01127                     AGAAGTTAGAAGAGTGAGCAGGAATAGTCTTGTGAATGGGGCCACGATCGTGCAACTTGCAGAGCCTTTCGCTGATCTAGCAATACGGTATTTGAAATCG
BetulapubescensGer407526-B5_TA   AGAAGTTAGAAGAGTGAGCAGGAATAGTCTTGTGAATGGGGCCACGATCGTGCAACTTGCAGAGCCTTTCGCTGATCTAGCAATACGGTATTTGAAATCG

                                 510       520       530       540       550       560       570       580       590       600
                                 ....|....|....|....|....|....|....|....|....|....|....|....|....|....|....|....|....|....|....|....|
contig_03284
contig_01127                     ACTGGGGAAGAAACAAGACTGTTCAAAAAGTTTCCTCGTGATTTATCGCCGGCGAAGCATCTCGCGTTCGATTTGTTTCTGGGGTCGATTTTGGTACAT
BetulapubescensGer407526-B5_TA   ACTGGGGAAGAAACAAGACTGTTCAAAAAGTTTCCTCGTGATTTATCGCCGGCGAAGCATCTCGCGTTCGATTTGTTTCTGGGGTCGATTTTGGTACAT

                                 610       620       630       640       650       660       670       680       690       700
                                 ....|....|....|....|....|....|....|....|....|....|....|....|....|....|....|....|....|....|....|....|
contig_03284
contig_01127                     TAAACGCGCAAGAGA---
BetulapubescensGer407526-B5_TA   TAAACGCGCAAGAGAAGATGGTGATTCAGATTCTAAACACTAGAATATTCAAAACAGAAGATCAAGATCGTCAAAGATGGCAAATCGTCCAATAATTGT

                                 710       720       730       740       750       760       770       780       790       800
                                 ....|....|....|....|....|....|....|....|....|....|....|....|....|....|....|....|....|....|....|....|
contig_03284
contig_01127
BetulapubescensGer407526-B5_TA   TCAAGTAATTAAGCCACCGATATTGGCAAATGTTCAAATGATAGTTCAAGCAAAGCCAGTGATGGCAAATCGTCAAATGATGATTCAAGCAAAGCCAAGA

                                 810       820
                                 ....|....|....|....|.
contig_03284
contig_01127
BetulapubescensGer407526-B5_TA   AGAAACGCTTGAGGCTTGAGT
```

Figure A2 Sanger sequence alignment of Capillovirus-like sequence with contig 03284. The random evaluation of the sequence environment around the integrated capillovirus coat protein in **FXXK01000935.1 contig934** *Betula pendula* revealed the existence of RdRP, RdRP, transposon Ty3-IGagPol polyprotein, Fbox RNase, gag protein, integrase and gag pol polyprotein. In addition, it became clear that the capillovirus coat protein sequence was integrated several times (2-3 times in Contig934).

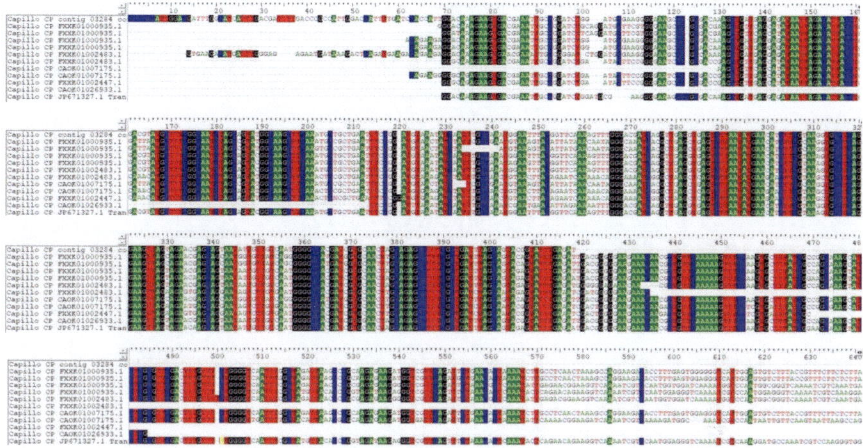

Figure A3 Sanger sequence alignment of Capillovirus-like sequence with contig 03284. The random evaluation of the sequence environment around the integrated capillovirus coat protein in FXXK01000935.1 contig934 *Betula pendula* revealed the existence of RdRP, RdRP, transposon Ty3-IGagPol polyprotein, Fbox RNase, gag protein, integrase and gag pol polyprotein. In addition, it became clear that the capillovirus coat protein sequence was integrated several times (2-3 times in Contig934).

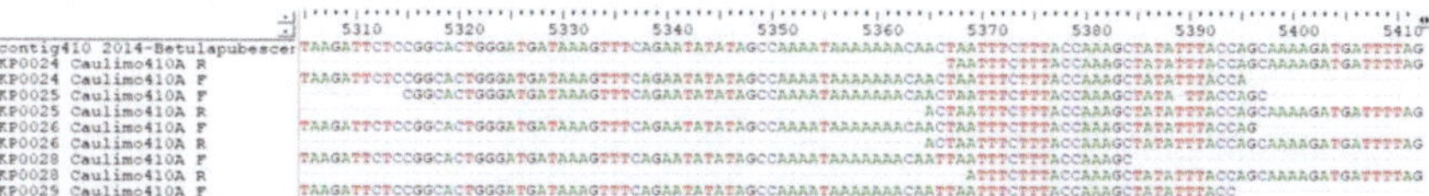

Figure A4. Confirmation of partial fragment of Birch Caulimovirus sequence which was amplified with primer pair combination of (Caulimo410F/-R) on nucleotide level using databases (contig410 from 2014 tree in Finland BetulapubescensFin407501-3A; contig1079 from 2014 tree in Finland BetulapubescensFin407501_3A) as reference sequence and 5 RT-PCR product (KP0024, KP0025, KP0026, KP0027, KP0028, KP0029).

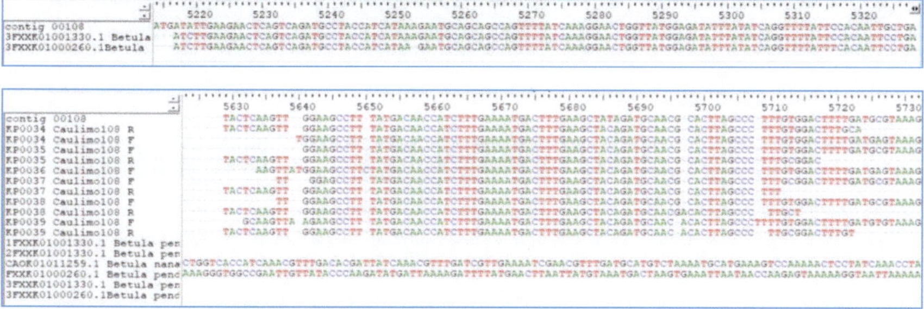

Figure A5 The sequence alignment from the NCBI databank showing partial integration of Caulimovirus contig 108 with *Betula pendula* genome assembly (FXXK01001330.1 and 3FXXK01000260 with contig 1329 and contig 259 respectively.

```
RID: X7XEF5ZN015
Job Title:H0203_Bet_Finland_2016a 1,737 reads from H0203_Paired...
Program: BLASTN
Query: H0203_Bet_Finland_2016a 1,737 reads from H0203_Paired Reads ( mapped to BLRaV_Bet_Finland_2014b_AR (NC_040635) using Geneious ID: lcl|Query_44427(dna)
Database: wgs (2 databases)


Sequences producing significant alignments:
                                                          Max   Total Query  E     Per.
Description                                               Score Score cover Value Ident Accession
Betula nana WGS project CAOK00000000 data, contig 62091, whole... 86.9  86.9  1%  9e-14 76.61  CAOK01062084.1
Betula nana WGS project CAOK00000000 data, contig 4687, whole...  86.9  86.9  1%  9e-14 76.61  CAOK01004687.1
Betula pendula genome assembly, contig: Contig519, whole genom... 63.5  172   0%  1e-06 87.04  FXXK01000520.1
Betula nana WGS project CAOK00000000 data, contig 340, whole...   63.5  63.5  0%  1e-06 84.21  CAOK01000340.1
Betula nana WGS project CAOK00000000 data, contig 17115, whole... 59.9  59.9  0%  1e-05 83.64  CAOK01017113.1
Betula nana WGS project CAOK00000000 data, contig 8411, whole...  59.9  59.9  0%  1e-05 78.57  CAOK01008411.1
Betula pendula genome assembly, contig: Contig1725, whole geno... 59.0  158   0%  1e-05 85.19  FXXK01001726.1
Betula pendula genome assembly, contig: Contig1329, whole geno... 59.0  59.0  1%  1e-05 71.64  FXXK01001330.1
Betula pendula genome assembly, contig: Contig1188, whole geno... 59.0  109   1%  1e-05 71.64  FXXK01001189.1
Betula pendula genome assembly, contig: Contig2691, whole geno... 56.3  110   1%  2e-04 70.63  FXXK01002692.1
```

Figure A6. Some results from Blastn BLRaV against whole genome shotgun with *Betula (taxid 3504)* with different Maximum score, query cover, E-value, percentage identity as well as Accession numbers.

```
Betula pendula ( taxid 3505) - Notepad                                                                        —  □
File  Edit  Format  View  Help
RID: DVRRAEZD016
Job Title:H0005_Bet_Germany_2014a_BLRaV 2,682 reads...
Program: BLASTN
Query: H0005_Bet_Germany_2014a_BLRaV 2,682 reads from H0005_Paired Reads ( mapped to BLRaV_Bet_Finland_2014b_AR (NC_040635) using Geneious ID: lcl|Query_64
Database: wgs (WGS_VDB://FXXK01)


Sequences producing significant alignments:
                                                          Max   Total Query  E     Per.
Description                                               Score Score cover Value Ident Accession
Betula pendula genome assembly, contig: Contig1177, whole geno... 59.9  59.9  0%  6e-06 78.57  FXXK01001178.1
Betula pendula genome assembly, contig: Contig2241, whole geno... 58.1  58.1  2%  2e-05 66.09  FXXK01002242.1
Betula pendula genome assembly, contig: Contig1422, whole geno... 52.7  52.7  1%  9e-04 70.07  FXXK01001423.1
Betula pendula genome assembly, contig: Contig939, whole genom... 52.7  102   1%  9e-04 70.75  FXXK01000940.1
Betula pendula genome assembly, contig: Contig53, whole genome... 52.7  93.6  0%  9e-04 84.00  FXXK01000054.1
Betula pendula genome assembly, contig: Contig2802, whole geno... 51.8  51.8  0%  9e-04 84.00  FXXK01002803.1
Betula pendula genome assembly, contig: Contig507, whole genom... 51.8  189   1%  9e-04 84.00  FXXK01000508.1
Betula pendula genome assembly, contig: Contig2094, whole geno... 50.9  96.3  0%  0.003 82.69  FXXK01002095.1
Betula pendula genome assembly, contig: Contig1188, whole geno... 50.9  100   1%  0.003 69.17  FXXK01001189.1
Betula pendula genome assembly, contig: Contig918, whole genom... 50.9  50.9  1%  0.003 69.44  FXXK01000919.1
Betula pendula genome assembly, contig: Contig876, whole genom... 50.9  50.9  1%  0.003 69.47  FXXK01000877.1
Betula pendula genome assembly, contig: Contig4154, whole geno... 50.0  50.0  0%  0.003 86.36  FXXK01004155.1
Betula pendula genome assembly, contig: Contig4139, whole geno... 50.0  50.0  0%  0.003 86.36  FXXK01004140.1
Betula pendula genome assembly, contig: Contig4017, whole geno... 50.0  50.0  0%  0.003 86.36  FXXK01004018.1
Betula pendula genome assembly, contig: Contig3498, whole geno... 50.0  50.0  0%  0.003 86.36  FXXK01003499.1
Betula pendula genome assembly, contig: Contig3480, whole geno... 50.0  50.0  0%  0.003 86.36  FXXK01003481.1
Betula pendula genome assembly, contig: Contig3004, whole geno... 50.0  139   0%  0.003 86.36  FXXK01003005.1
Betula pendula genome assembly, contig: Contig2845, whole geno... 50.0  50.0  0%  0.003 86.36  FXXK01002846.1
Betula pendula genome assembly, contig: Contig2832, whole geno... 50.0  50.0  0%  0.003 86.36  FXXK01002833.1
Betula pendula genome assembly, contig: Contig2788, whole geno... 50.0  50.0  0%  0.003 86.36  FXXK01002789.1
Betula pendula genome assembly, contig: Contig2786, whole geno... 50.0  50.0  0%  0.003 86.36  FXXK01002787.1
Betula pendula genome assembly, contig: Contig2765, whole geno... 50.0  50.0  0%  0.003 86.36  FXXK01002766.1
Betula pendula genome assembly, contig: Contig2606, whole geno... 50.0  50.0  0%  0.003 86.36  FXXK01002607.1
Betula pendula genome assembly, contig: Contig2605, whole geno... 50.0  50.0  0%  0.003 86.36  FXXK01002606.1
```

Figure A7. Some results from Blastn BLRaV against whole genome shotgun with *Betula pendula (taxid 3505) producing* significant sequence alignments with different Maximum score, query cover, E-value, percentage identity and accession numbers.

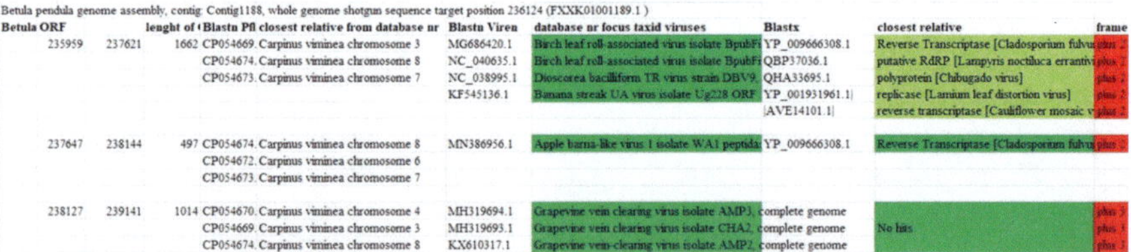

Betula pendula genome assembly, contig: Contig1188, whole genome shotgun sequence target position 236124 (FXXK01001189.1)

Betula ORF		lenght of	Blastn Pfl closest relative from database nr	Blastn Viren	database nr focus taxid viruses	Blastx	closest relative	frame
235959	237621	1662	CP054669. Carpinus viminea chromosome 3	MG686420.1	Birch leaf roll-associated virus isolate BpubFi	YP_009666308.1	Reverse Transcriptase [Cladosporium fulvu...	plus 2
			CP054674. Carpinus viminea chromosome 8	NC_040635.1	Birch leaf roll-associated virus isolate BpubFi	QBP37036.1	putative RdRP [Lampyris noctiluca errantiv...	plus 2
			CP054673. Carpinus viminea chromosome 7	NC_038995.1	Dioscorea bacilliform TR virus strain DBV9	QHA33695.1	polyprotein [Chibugado virus]	plus 2
				KF545136.1	Banana streak UA virus isolate Ug228 ORF	YP_001931961.1|	replicase [Lamium leaf distortion virus]	plus 2
						|AVE14101.1|	reverse transcriptase [Cauliflower mosaic v...	plus 2
237647	238144	497	CP054674. Carpinus viminea chromosome 8	MN386956.1	Apple barna-like virus 1 isolate WA1 peptida	YP_009666308.1	Reverse Transcriptase [Cladosporium fulvu...	plus 2
			CP054672. Carpinus viminea chromosome 6					
			CP054673. Carpinus viminea chromosome 7					
238127	239141	1014	CP054670. Carpinus viminea chromosome 4	MH319694.1	Grapevine vein clearing virus isolate AMP3, complete genome			plus 3
			CP054669. Carpinus viminea chromosome 3	MH319693.1	Grapevine vein clearing virus isolate CHA2, complete genome		No hits	plus 3
			CP054674. Carpinus viminea chromosome 8	KX610317.1	Grapevine vein-clearing virus isolate AMP2, complete genome			plus 3

Figure A8: The bioinformatic analysis of the *Betula* sp. after BLASTx of BLRaV genome (*H0005 Bet-Germany-2014a-BLRaV 2,682 reads from H0005-Paired Reads mapped to BLRaV-Bet-Finland-2014b-AR (NC-040635)*) with *Betula pendula* genome, contig 1188, whole genome shotgun sequence position 236124 (FXXK01001189.1) against the database whole genome shotgun sequence (WGS) resulted in the extraction of partial sequences belonging to the group of badnavirus similar to BLRaV.

12 Acknowledgement

I am very grateful for the support and help in any regard during the time I spent at the division of Phytomedicine-Humboldt University of Berlin. Special thanks go to Prof. Dr. Carmen Büttner for her great support. I want to thank Dr. Maria Landgraf for her great supervision through out my entire PhD. Thank you very much for supporting the bioinformatical analyses and scientific discussion.

I thank Ms. Andrea Klinke for her kind assistance during my laboratory experiments. Thank you very much to Dr. Susanne von Bargen and Dr. Martina Bandte for also providing me leaf samples from Finland. I thank the pollenPal project group especially Prof. Dr. Susanne Jochner-Oette for providing me with sample materials used in this research. Special thanks go to the 'Grünflächenamt' for providing us with lifts trucks and driving us around streets of Berlin during sample collection. I would like to thank the entire phytomedicine group for their kind support during my stay at the division.

Finally, I would like to thank my family for their support and love through out the struggles during PhD.

13 Declaration on independence and resources

I hereby declare that this thesis is the result of my own work which has been written only with the help of the indicated sources and that this thesis, in same or similar form has not been submitted to any institution yet.

Date: 20.09.2022

Elisha Bright Opoku

In der Reihe *Berliner ökophysiologische und phytomedizinische Schriften* **sind bisher erschienen:**

Band 01:   Mohammad Mahir Uddin (2009)
             Chemical ecology of mustard leaf beetle Phaedon cochleariae (F.).
             ISBN 978-3-89959-848-3.

Band 02:   Ilir Morina (2009)
             Entwicklung von Verfahren zur Rekultivierung der Aschedeponie des
             Braunkohlekraftwerks in Prishtina (Kosovo).
             ISBN 978-3-89959-872-8.

Band 03:   Melanie Wiesner (2009)
             Veränderungen gesundheitsrelevanter Inhaltsstoffe in *Parthenium
             hysterophorus* L. in Abhängigkeit von der Pflanzengröße und Klimafaktoren.
             ISBN 978-3-89959-880-3.

Band 04:   Fransika Rohr (2009)
             Variabilität aliphatischer Glucosinolate in *Arabidopsis thaliana*-Ökotypen und
             deren Einfluss auf die Wirtspflanzeneignung von zwei folivoren Insektenarten.
             ISBN 978-3-89959-884-9.

Band 05:   Jutta Buchhop (2009)
             Characterization of phylogenetically diverse CLRV-isolates by RFLP and
             research into identification of two isometric viruses.
             ISBN 978-3-89959-929-9.

Band 06:   Nora Koim (2010)
             Urban sprawl, land cover change and forest fragmentation – Case study
             Pereira, Colombia.
             ISBN 978-3-89959-955-8.

Band 07:   Nadja Förster (2010)
             Eignung unterschiedlicher salicylathaltiger *Salix*-Klone für die
             Arzneimittelindustrie.
             ISBN 978-3-89959-964-0.

Band 08:   Jana Gentkow (2010)
             Cherry leaf roll virus (CLRV): Charakterisierung ausgewählter Virusisolate
             unter besonderer Berücksichtigung des viralen Hüllproteins.
             ISBN 978-3-89959-976-3.

Band 09:   Ahmad Fakhro (2010)
             Interaction of Pepino mosaic virus (PepMV) and fungal root endophytes with
             tomato hosts (*Lycopersicum esculentum* Mill.).
             ISBN 978-3-89959-995-4.

Band 10:   Stefan Irrgang (2010)
             Mikro- und makroskopische Untersuchungen an Veredelungsstellen von
             Straßenbäumen im Hinblick auf die Beeinflussung ihrer Bruchsicherheit.
             ISBN 978-3-89959-998-5.

Band 11:   Julia Jahnke (2010)
             Guerilla Gardening anhand von Beispielen in New York, London und Berlin.
             ISBN 978-3-86247-001-3.

Band 12:   Astrid Karoline Günther (2010)
            Analysen zur Intensität der Pflanzenschutzmittel-Anwendung und Aufklärung
            ihrer Einflussfaktoren in ausgewählten Ackerbaubetrieben.
            ISBN 978-3-86247-005-1.

Band 13:   Milena A. Dimova (2010)
            Untersuchungen zur Epidemiologie von *Pythium aphanidermatum* in
            Abhängigkeit von den Umgebungsbedingungen bei der Gewächshausgurke
            (*Cucumis sativus* L.).
            ISBN 978-3-86247-033-4.

Band 14:   Claudia Patricia Pérez-Rodríguez (2010)
            Physiologische Veränderungen in Früchten der Solanaceaengewächse in
            Abhängigkeit von physikalischen Elicitoren während der Produktion und nach
            der Ernte.
            ISBN 978-3-86247-066-2.

Band 15:   Charles Adarkwah (2010)
            Integrated management of the stored-product pest insects *Corcyra cephalonica*,
            Cadra cautella, *Sitophilus zeamais* and *Tribolium castaneum* by use of the
            parasitic wasps *Habrobracon hebetor*, *Venturia canescens*, *Lariophagus
            distinguendus* and neem seed oil.
            ISBN 978-3-86247-077-8.

Band 16:   Christoph von Studzinski (2010)
            Angewandte Methoden der xenovegetativen Vermehrung.
            ISBN 978-3-86247-088-4.

Band 17:   Tanja Mucha-Pelzer (2011)
            Amorphe Silikate – Möglichkeiten des Einsatzes im Gartenbau zur
            physikalischen Schädlingsbekämpfung.
            ISBN 978- 3-86247-106-5.

Band 18:   Diego Miranda (2011)
            Effect of salt stress on physiological parameters of cape gooseberry, *Physalis
            peruviana* L.
            ISBN 978- 3-86247-119-5

Band 19:   Franziska Beran (2011)
            Host preference and aggregation behavior of the striped flea beetle, *Phyllotreta
            striolata*.
            ISBN 978- 3-86247-188-1

Band 20:   Mohammed Abul Monjur Khan (2011)
            Induced biochemical changes and gene expression in *Brassica oleracea* and
            Arabidopsis thaliana by drought stress and its consequences on resistance to
            aphids.
            ISBN 978- 3-86247-203-1.

Band 21:   Sandra Lerche (2012)
            Untersuchungen zur Anwendung, Praxiseinführung und molekularen
            Identifizierung von Stamm V24 des entomopathogenen Pilzes *Lecanicillium
            muscarium* (Petch) Zare & W. Gams.
            ISBN 978- 3-86247-248-2.

Band 22:   Carsten Richter (2012)
              Entwicklung und Überprüfung eines gasdichten Küvettensystems für
              Experimente unter hochgradig kontrollierten Bedingungen mit
              Gaswechselmessungen.
              ISBN 978- 3-86247-271-0.

Band 23:   Aksana Grineva (2012)
              Influence of the two stored grain pest insects *Sitophilus granarius* and
              Oryzaephilus surinamensis on temperature, relative humidity, moisture
              content, and mould growth in stored triticale.
              ISBN 978- 3-86247-279-6.

Band 24:   Carmen Büttner & Christian Ulrichs (2012)
              Aktuelle Themen in Landwirtschaft und Gartenbau am Beispiel von Südtirol.
              ISBN 978- 3-86247-279-6.

Band 25:   Juliane Langer (2012)
              Molecular and epidemiological characterisation of Cherry leaf roll virus
              (CLRV).
              ISBN 978- 3-86247-279-6.

Band 26:   Franziska Rohr-Doucet (2012)
              AOP-Variabilität in *Arabidopsis thaliana*-Kreuzungslinien – Auswirkungen
              auf die Resistenz gegenüber verschieden spezialisierten Lepidopteren-Arten.
              ISBN 978- 3-86247-329-8.

Band 27:   Vanessa Hörmann (2012)
              Lignin als biologische Barriere gegen Schimmelpize in Innenräumen.
              ISBN 978- 3-86247-330-4.

Band 28:   Jacqueline Kurth (2013)
              Auswirkungen verschiedener Düngerzusammensetzungen auf den Ertrag bei
              Schnittrosen unter Berücksichtigung des Anbauverfahrens.
              ISBN 978- 3-86247-336-6.

Band 29:   Juliane Langer, Carmen Büttner & Christian Ulrichs (2014)
              Kolumbien – klimatische und politische Voraussetzungen für eine
              landwirtschaftliche Produktion.
              ISBN 978- 3-86247-430-1.

Band 30:   Heike Luisa Dieckmann (2014)
              Detection of the European mountain ash ringspot associated virus (EMARaV)
              in Sorbus aucuparia L. in several European contries.
              ISBN 978- 3-86247-441-7.

Band 31:   Rima Marion Baag (2014)
              Analyse von trans-Resveratrol in historischen Rebsorten der Weinanbaugebiete
              Sachsen und Saale-Unstrut.
              ISBN 978- 3-86247-488-2.

Band 32:   Ayesha Rahmann (2014)
              Study of the protective effects of nano-structured silica and plant derived
              biomolecules on nuclear polyhedrosis virus affected silkworm larvae at the
              behavioral and molecular level.
              ISBN 978- 3-86247-495-0.

Band 33:   Bettina Gramberg (2015)
            Weiterentwicklung eines elektrochemischen Biosensors zum Nachweis von
            Pflanzenviren und Insektiziden.
            ISBN 978- 3-86247-512-4.

Band 34:   Wilhelm van Husen (2015)
            Artspezifische Aufnahme und Verteilung von Cadmium bei indigenen
            afrikanischen Gemüsearten und daraus abzuleitende Ernährungsempfehlungen.
            ISBN 978- 3-86247-523-0.

Band 35:   Jenny Roßbach (2015)
            European mountain ash ringspot-associated viras (EMARaV): diversity and
            geographic distribution in Europe.
            ISBN 978- 3-86247-547-6.

Band 36:   Silke Steinmöller (2015)
            Risikominderung der Verbreitung von Quarantäneschadorganismen der
            Kartoffel durch hygienisierende Maßnahmen.
            ISBN 978- 3-86247-550-6.

Band 37:   Christin Siewert (2016)
            Genomic and functional analysis of species within the Acholeplasmataceae –
            Phytoplasmas and Acholeplasmas.
            ISBN 978- 3-86247-579-7.

Band 38:   Angela Köhler (2016)
            Untersuchungen zur Phenolglycosidkonzentration ausgewählter intra- und
            interspezifischer Kreuzungen salicinreicher Biomasseweiden.
            ISBN 978- 3-86247-581-0.

Band 39:   Nicolas Meyer (2016)
            Vergleichende ökophysiologische Untersuchung verschiedener Baumarten zur
            Verwendung als Straßenbegleitgrün in Berlin.
            ISBN 978- 3-86247-586-5.

Band 40:   Stefanie Schläger (2017)
            Identification of variation within sex pheromone blends of various Maruca
            vitrata populations for refining pheromone lures and traps in Asia.
            ISBN 978- 3-7369-9570-3.

Band 41:   Elisha Otieno Gogo (2017)
            Pre- and postharvest treatments for the quality assurance of African indigenous
            leafy vegetables.
            ISBN 978-3-7369-9650-2.

Band 42:   Luise Dierker (2017)
            Interaktion des RNA2-kodierten Transportproteins (MP) des *Cherry leaf roll
            virus* (CLRV) mit dem viralen Hüllprotein (CP) und pflanzlichen
            Wirtsfaktoren
            ISBN 978-3-7369-9670-0.

Band 43:   Nadja Förster (2017)
            Antikarzinogenes Potential ausgewählter Glucosinolate von *Moringa oleifera*
            ISBN 978-3-7369-9704-2.

Band 44:   Steffen Pallarz (2018)
             Data driven classification of host-plant response (virus-plant)
             ISBN 978-3-7369-9731-8.

Band 45:   Vanessa Hörmann (2018)
             Biofiltration of indoor pollutants by ornamental plants
             ISBN 978-3-7369-9815-5.

Band 46:   Allan Ndua Mweke (2018)
             Development of entomopathogenic fungi as biopesticides for the management
             of Cowpea Aphid, *Aphis craccivora* Koch
             ISBN 978-3-7369-9908-4.

Band 47:   Isabella Linda Bisutti (2019)
             Biological agents formulation and mode of application against strawberry
             diseases
             ISBN 978-3-7369-7032-8.

Band 48:   Judith Henze (2019)
             Innovation in Agriculture: The Potential, Challenges and Adoption and
             Diffusion of Aquaponics and Agricultural Mobile Phone Application in Kenya
             ISBN 978-3-7369-7133-2.

Band 49:   Maliha Gul Aftab (2021)
             Chemical ecology of Cabbage White (Pieris sp.)
             ISBN 978-3-7369-7498-2.

Band 50:   Simon Goisser (2021)
             Suitability of portable NIR sensors (food-scanners) for the determination of
             fruit quality along the supply chain using the example of tomatoes
             ISBN 978-3-7369-7543-9.

Band 51:   Guido Rux (2022)
             Fresh-cut apples: Aspects of respiration, sanitation and storage conditions
             affecting quality and volatile synthesis
             ISBN 978-3-7369-7564-4.

Band 52:   Joseph Cutler (2022)
             New strategies to ensure food production: Certification of virus-tested plant
             material in Colombia
             ISBN 978-3-7369-7603-0.